THE BIG BOOK OF MARS

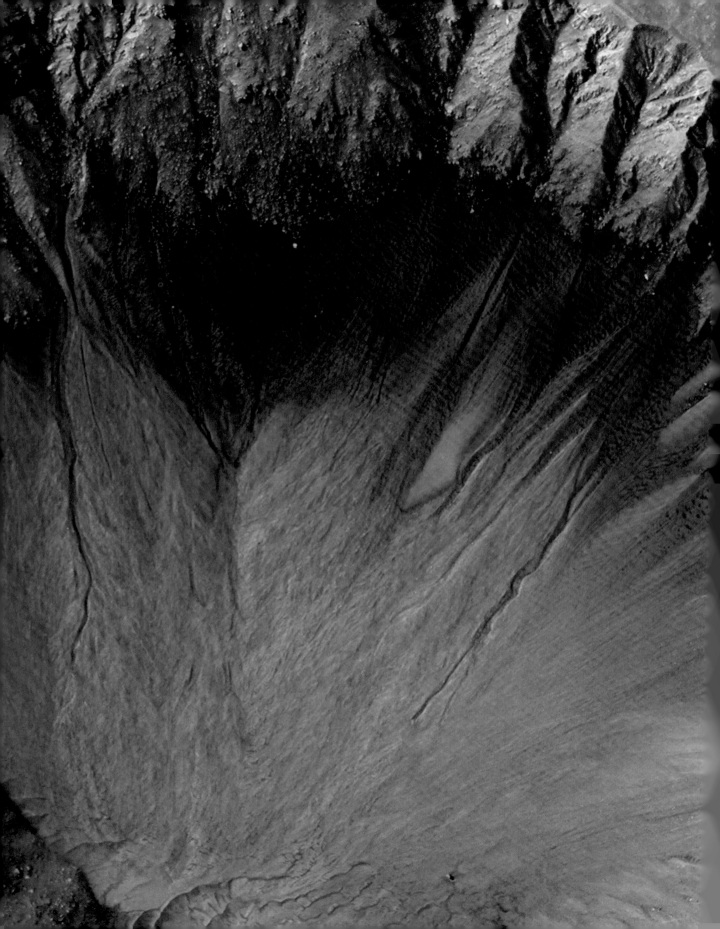

THE BIG BOOK OF MARS

From Ancient Egypt to *The Martian*, A Deep-Space Dive into Our Obsession with the Red Planet

MARC HARTZMAN

QUIRK BOOKS
PHILADELPHIA

Copyright © 2020 by Marc Hartzman

All rights reserved. Except as authorized under U.S. copyright law, no part of this book may be reproduced in any form without written permission from the publisher.

Full Library of Congress Cataloging in Publication Data available upon request.

ISBN: 978-168369-209-6

Printed in China

Typeset in Sentinel, ITC Avant Garde, and House Gothic

Cover and interior designed by Ryan Hayes
Cover illustration by The Brave Union
Full photo credits appear on page 252.
Production management by John J. McGurk

Quirk Books
215 Church Street
Philadelphia, PA 19106
quirkbooks.com

10 9 8 7 6 5 4 3 2 1

TO LIZ, LELA, AND SCARLETT,
MY THREE FAVORITE EARTHLINGS

AND OOMARURU, MY FAVORITE MARTIAN

CONTENTS

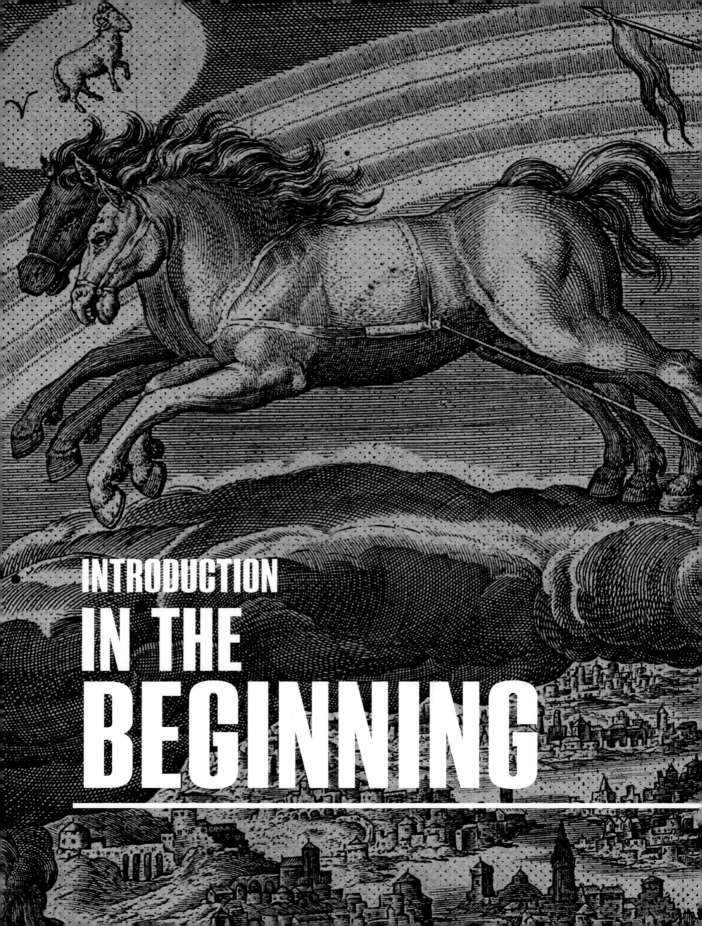

INTRODUCTION
IN THE
BEGINNING

One upon a time, the planet Mars was a wondrous place filled with intelligent beings. Standing ten to twenty feet tall, these brilliant Martians feverishly dug canals to irrigate their dry planet and desperately tried to contact us, the people of Earth.

Some hundred years ago, that's what the greatest minds in the world believed, and so it was. This was an era when telephones, automobiles, airplanes, radios, and television were new and magical wonders. The impossible, it seemed, was never more possible. So, when newspapers spread scientific theories and hyped the next big advance—communication with Martians—people thought, why not? The long-distance phone bills would cost a fortune, but still, what a time to be alive!

I discovered this peculiar period of interplanetary excitement while researching Nikola Tesla's attempts to communicate with our supposed neighbors. In the process I stumbled upon someone who'd claimed to already be hobnobbing with them. His name was Dr. Hugh Mansfield Robinson, and in the 1920s he was in touch with a big-eared, nearly seven-foot-tall Martian woman named Oomaruru. Telepathically.

Forget Tesla. Who was this Robinson guy?

I switched my focus and dug into this other story. For someone who likes writing about weird history, this was striking gold. Robinson's conversations with Oomaruru made headlines all around the world. Earth was fascinated with all things Mars, regardless of who was offering it.

The obsession with the Red Planet goes back even further, to a time long before the early twentieth century. Thousands of years ago, at least, when its red color conjured images of bloodshed instead of the oxidized iron dust covering its surface. Ancient Egyptian astronomers noticed the fiery orb's retrograde motion and named it *Har Decher*, meaning "the Red One."

As time trudged forward, other curious civilizations gave it their own names based on its blazing visage. Chinese astronomers saw the "fire star" and considered it a sign of "bane, grief, war and murder." The Hebrews called it *Ma'adim*, from their word for red, *adom*. They believed that those born under the influence of Mars would enter bloody occupations, such as murder, the military, or surgery. By 600 BCE, the Babylonians dubbed it *Nergal*, after their god of war and destruction. Bloodshed became the presumptuous theme for the Red Planet. For whatever barbaric acts these peoples were committing in their own kingdoms, they must've thought all hell was breaking loose on faraway Mars.

When the Babylonians gazed at the skies to develop a calendar, they created the seven-day week and named each day for a known celestial body: the sun, the moon, Mars, Mercury, Venus, Jupiter, and Saturn. The day attributed to Mars is Tuesday (the association is more apparent in Spanish, in which the day is known as *Martes*);

its association with war and aggression led them to perform special rituals on Tuesdays to avoid hostile influence from the Red Planet. Thousands of years later, Mars remains in our lives at least once a week.

In the mid-300s BCE, while pondering, well, everything, Aristotle observed the moon passing in front of Mars, which led him to conclude that it, and the other planets, were farther away from Earth. In the meantime, he and the rest of Greek civilization carried on the tradition and named the Red Planet *Ares*, after their own god of war. The son of Zeus and Hera, Ares represented the unfortunate but necessary brutality and violence of battle. Ares's two sons, Phobos and Deimos (meaning "fear" and "terror," respectively), would lend their names to the two moons of Mars after their 1877 discovery by astronomer Asaph Hall.

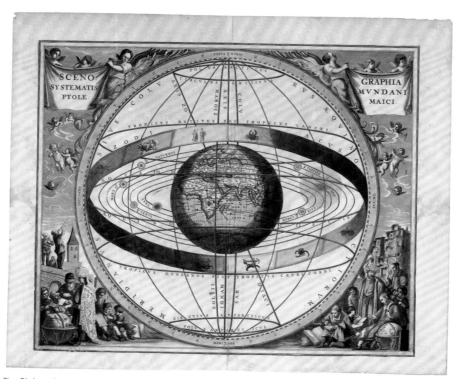

The Ptolemaic system showing the signs of the zodiac and the solar system, with Earth at the center.

Mars in His Chariot, 16th century.

All these gods of war eventually lost the planet-naming contest to the Roman deity, Mars. The planet's mystique mirrored their affinity for power and bloody paths of conquest. But as far as the Roman astronomer Claudius Ptolemy was concerned, Mars was just a soldier in the earthly universe. It, along with the sun and other celestial bodies, revolved around us. His theories, known as the Ptolemaic System, reigned for the next 1,400 years, until Copernicus proclaimed a different idea about the organization of the solar system. From then on, observations about Mars grew more detailed, and the obsession with what secrets lay across the cosmos grew exponentially. Were we alone in the universe, or were other civilizations staring right back at us wondering the same thing?

The Big Book of Mars examines the many enthusiastic answers offered by some of history's greatest and most creative minds—from Nikola Tesla and Guglielmo Marconi to H. G. Wells, Ray Bradbury, and many more, including a bunch of confident UFO spotters.

As I dug into the Dr. Mansfield Robinson story, I found stories about others who believed that communication with Martians was inevitable—and could happen any day. I suddenly found myself obsessed with their obsession, and I wanted to learn more about these theoretical Martians and where all the ideas about talking to them came from. My own journey was just beginning. I scoured newspaper archives and discovered countless articles, most speculating about the existence of canals on the surface of Mars. That fantastically wrong belief, born in the nineteenth century, inspired the imaginations of Wells, Edgar Rice Burroughs, and other early science-fiction writers. Their stories, and those of many others, inspired the next generation of scientists who were determined to invent the space travel their favorite writers had imagined. And so they did. Science leads to science fiction, and science fiction leads to science. It's a symbiotic relationship.

Like those geniuses who turned imaginations into reality, I devoured as many Mars-related science-fiction books as I could and watched a lot of movies (mostly bad ones), from *The Wizard of Mars* to *Santa Claus Conquers the Martians*. I also made a pilgrimage to the place where engineering feats happen all the time: the NASA Jet Propulsion Laboratory in Pasadena, California. The sprawling JPL campus is home to the dedicated teams who have been making journeys to Mars possible since the 1960s. The mission control room is surrounded by laboratories, testing areas, and the kinds of signs you just don't find on other tours, like "Mars Yard" and "Planetary Landing Testbed." Even though this close-knit bunch of men and women are busily writing the sci-fi movie that we're all living in, they know how to take a break from equations and algorithms. I know this, having witnessed a group of them playing their trumpets in a room adjacent to a robotics lab.

Smart as they are, they've yet to build a time machine, although visiting the Lowell Observatory in Flagstaff, Arizona, felt as though I'd stepped into one. Rummaging through Percival Lowell's personal scrapbooks, drawings, and notes from 1894 through the early 1900s, and looking through a telescope at Mars in the spot where Lowell so often did, was definitely a step back to an incredibly exciting time in Earth–Mars history. As was standing on the spot in Grover's Mill, New Jersey, where the Martians began their invasion on October 30, 1938, during Orson Welles's radio broadcast of *The War of the Worlds*. The town's commemoration of the fictional attack has transformed the terrifying event into nostalgic pride.

These moments are just a small part of my treasure hunt through history, which included many conversations with the people who've helped pave the interplanetary road to Mars that sends us photos from its surface every day. (At one point, that idea would've sounded as crazy as the concept of telepathy with a giant Martian woman named Oomaruru.)

Assuming the space community reaches its goal of putting humans on Mars by the 2030s, the first person to take that historic step is already among us. To that person I say, may this book serve as a guide through humanity's history with the Red Planet—until *you* write the next chapter.

CHAPTER 1
BIG HOPES AND BIGGER
MISUNDERSTANDINGS

I n 1877 the universe as we know it changed and life on Earth was no longer the center of it. That year, Italian astronomer Giovanni Schiaparelli discovered something strange about Mars. The planet appeared to be covered with a peculiar network of linear structures. He called these lines *canali*, meaning "channels," but the word was translated into English as "canals." And if there were canals, then there must be living beings to dig them.

At the time, the completion of the Suez Canal was a relatively recent achievement that had required ten years of painstaking excavation. This marvel of human engineering seemed like bupkes compared to the Martians covering an entire planet with such extraordinary feats of labor. To do so, they must have all gotten along and worked well as a team. Clearly, they were far more advanced than we are. Still, Schiaparelli wasn't convinced that the *canali* were artificially made, but he didn't discount it, either.[1]

It took a few years until others began seeing the lines Schiaparelli saw, but once they did, many believed they had all the irrefutable proof they needed of extraterrestrial life. French astronomer Camille Flammarion was among the earliest and most vocal supporters of such beliefs. In fact, he'd been saying Mars was inhabited for years. His first book, *La pluralité des mondes habités* (*The Plurality of Inhabited Worlds*) was published in 1862, just three years after Darwin's *On the Origin of Species* introduced the concept of natural selection and evolution. In his writings, Flammarion argued for the existence of extraterrestrial life and imagined that just like life on Earth, it would simply adapt to its environment. The laws of evolution, he reasoned, could apply anywhere in the physical universe.

---

1 Then again, he also didn't discount the psychic powers of Italian medium Eusapia Palladino. In 1892, Schiaparelli and other scientists conducted a series of experiments with Palladino over the course of seventeen sittings. The group found it "difficult to attribute the phenomena produced to deception" but could not conclusively say there was no fraud.

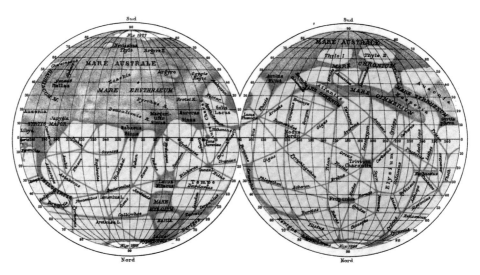

Giovanni Schiaparelli's atlas of Mars, 1888. Note that the South Pole is on top.

Of course, he was influenced by many thinkers long before Darwin. Mars has been quite the popular planet since the birth of modern astronomy. Toward the end of the sixteenth century, Danish astronomer Tycho Brahe closely observed the mysterious red star and accurately mapped its positions. His assistant, Johannes Kepler, correctly defined Mars's elliptical orbit and squashed the notion that all celestial bodies moved in circular paths. Their eyeballs served science well, but Galileo Galilei was one of the earliest early adopters: in 1610 he gazed at Mars through a newfangled gizmo called a telescope. This beta version didn't let him see any surface details, but what he did see was something solid: a clear disk as opposed to a shining star. Galileo also noticed that Venus showed phases like the moon, which supported Nicolaus Copernicus's 1543 theory that the planets revolved around the sun. Looking farther, he spotted four moons revolving around Jupiter, which offered further proof that not everything revolved around Earth. The Roman Church was no fan of Copernicus for suggesting that Earth wasn't the center of the universe, and these heliocentric and heretical observations only annoyed the holy powers that be even more. Galileo eventually ended up under house arrest, but his discoveries through the telescope kicked off the science of modern astronomy.

Portrait of Galileo, 1636.

As the seventeenth century trudged forward, Italian astronomer Francesco Fontana peered through a more powerful telescope and became the first man to draw Mars based on his observations. Christiaan Huygens topped Fontana and illustrated a map, complete with terrain features. By 1659, he had estimated the Martian day to be approximately twenty-four hours. Giovanni Cassini followed Huygens's lead and added forty minutes to the day (the Martian day, called a "sol," is now known to last twenty-four hours and thirty-seven minutes). He observed the southern polar ice cap to boot. Slowly but surely, we were getting to know the real Mars.

As details continued to emerge, curious minds began to ponder what life on the neighboring planet might look like. In his 1686 book, *Conversations on the Plurality of Worlds* (*right*), French essayist Bernard Le Bovier de Fontenelle described Mars as having high rocks that store up daylight and create "glorious illuminations" at night. "Great flocks" of luminous birds might also light up the Martian nights.

Page from *Conversations on the Plurality of Worlds*, 1780.

Years after mapping Mars, Huygens offered his thoughts on extraterrestrials, too. In the posthumously published *Cosmotheros; Or, Conjectures Concerning the Planetary Worlds, and Their Inhabitants,* he suggested that all the planets were likely inhabited by animals of some sort, with one species being superior.

"That there is some such rational Creature in the other Planets, which is the Head and Sovereign of the rest, is very reasonable to believe," he wrote, "for otherwise, were many endued with the same Wisdom and Cunning, we should have them always doing mischief, always quarrelling and fighting one another for Empire and Sovereignty." And to support those human-like creatures and other animals, surely there would be plants to nourish them. Huygens couldn't imagine that a divine creator would have it any other way.

By the eighteenth century, British astronomer Sir William Herschel set his telescope on Mars and took an even closer look than his predecessors. Herschel, who is best remembered for his discovery of Uranus in 1781,[2] began making his own optical instruments in the 1770s and soon surpassed the capabilities of any others. He hoped his creations would help prove his belief that the planets were inhabited—as was the sun, which he thought was populated by "beings whose organs are adapted to the peculiar circumstances of that vast globe."

"View of Dr. Herschel's forty-foot telescope," 1800s.

In 1779, Mars was in opposition, meaning that its orbital position is closest to Earth (with the sun being on the opposite side of us). This presented Herschel with a golden opportunity to test out his latest lenses and spot some Martians. But even with Mars's relative proximity, Herschel didn't find life. He did, however, notice dark spots and determined that they must be oceans. Those, along with perceived clouds, indicated a moderate atmosphere. And if that were the case, Herschel posited that Martians "probably enjoy a situation in many respects similar to our own."

2 Don't blame him for the name. He named it Georgium Sidus, or George's Star, to honor and please King George III. Years later, German astronomer Johann Elert Bode changed it to be consistent with the mythological naming convention of the other planets. Uranus was the ancient Greek god of the Heavens.

An Early
Martian Census

In 1838, Scottish minister and amateur astronomer Thomas Dick took the idea of extra-terrestrial life to another level when he decided to estimate the population of the entire universe. In his book, *Celestial Scenery; Or, the Wonders of the Planetary System Displayed*, he counts all the aliens by extrapolating from the population of Great Britain: 280 people per square mile. With a little fancy math and a plethora of planetary data, he then took a census of every celestial body, including the sun (which he estimated had 681,184,000,000,000 sun bathers).

Despite its smaller size, Mars would have "six millions of square miles more than on all the habitable parts of our globe" and therefore "would contain a population of more than fifteen thousand five hundred millions, which is nineteen times the number of the inhabitants of the earth; but, as it is probably that one third of the surface of Mars is covered with water, should we subtract one third from these sums there would still remain accommodation for twelve times the number of the population of our globe." That amounts to more than ten billion Martians. And one crowded planet.

f Herschel was right, how could we reach out to these beings who might be similar to ourselves? Early nineteenth-century scientists started brainstorming ideas. German mathematician and physicist Carl Friedrich Gauss was first to bat, proposing to communicate through the universal language of math. He wanted to create a massive right triangle and three squares—symbolizing the Pythagorean theorem—on the Siberian tundra using pine trees and fields of wheat. Surely such a simple display would impress any Martians and show off our smarts.

Austrian astronomer Joseph Johann von Littrow also thought large displays of basic shapes would fit the bill. Unlike Gauss's tundra tactics, von Littrow wanted to make a bigger spectacle by digging giant canals in the Sahara Desert and filling them with enough kerosene to burn for six hours at night. A circle, for example, with a twenty-mile diameter and trenches a few hundred yards wide would definitely catch Martians' eyes, however many they might have, and assure them that Earth abounded with intelligent life.

Unfortunately for the pyromaniacs and triangle aficionados of the time, both proposals went nowhere. By 1869, French inventor Charles Cros pitched a new scheme to his country's government. Rather than burn or build giant messages on Earth, why not do it on Mars? Now *that* would get the Martians' attention. Cros wanted to build a huge mirror that would direct the sun's rays at Mars and burn large letters right onto its surface by fusing its desert sand, like a sort of cosmic solar pen.

He'd start with simple shapes and then move on to more complex objects, like a house and a human. If defacing their planet wasn't the best way to say *Bonjour!*, Cros also suggested building multiple mirrors that, when strategically placed, could reflect sunlight in the shape of the Big Dipper toward Mars and offer evidence of our intelligence. The French government passed.[3]

All these early ideas and efforts were based purely on speculation and wonder, but by the late nineteenth century, Camille Flammarion (*below*) believed he had witnessed proof. The canals were working well—so well, in fact, that he believed the Red Planet's coloring came from its abundant vegetation. "Why, we may

---

3 In defense of Cros, not all his ideas were so out there. He's perhaps best remembered as the near-inventor of the phonograph. Cros submitted a written plan for the sound reproduction device to the Academy of Sciences shortly before Edison produced a working model. The two are not believed to have known about each other's ideas.

ask, is not the Martian vegetation green? Why should it be?—is the reply. From this point of view, there is no reason to regard the Earth as typical in the universe," Flammarion explained. His reasoning:

> "Moreover, the terrestrial vegetation can itself be reddish, and has been for the majority of the continents; the first terrestrial plants were lycopods, whose color is a 'Martian' reddish yellow. The green substance which gives our vegetation its color—chlorophyll—is made up of two elements; one green, the other yellow. These two elements can be separated by chemical processes. It is therefore perfectly scientific to admit that under conditions different from those on Earth, the yellow chlorophyll can exist alone, or be dominant."

As for the Martians, Flammarion thought they would not resemble humans because their life had formed under different conditions, but he felt certain their intelligence was superior to ours, for several reasons. "The first is that they could hardly be less intelligent than we are," he wrote in 1907, "seeing that we spend three-fourths of our resources and run heavily into debt simply to keep up armies and navies; and we cannot even agree upon a universal calendar or meridian." He continued:

> "The second reason is that progress is an absolute, irresistible law. If the inhabitants of Mars, as we have every reason to suppose, have gone through the regular process of slow development, their present condition ought to resemble what our own will be several million years hence, inasmuch as Mars is a much older planet than the Earth."

Since it was assumed that Martians were a much older and far more advanced species, Flammarion supposed they had tried communicating with us long ago, "when mammoths were roaming around our comparatively youthful planet." But since they never heard back, they likely concluded that "our astronomical science is only child's place besides theirs" and went on living their wonderfully harmonious lives.

As physical creatures, the astronomer offered the possibility that the lightness of their bodies may have led them to develop into a winged race. But he did not mean to imply they would look like birds. "The bats, are they not mammals which suckle their young?" he reasoned.

Perhaps Flammarion's strong beliefs in Martians were a reaction to his disappointment in humankind. He hoped there was something better out there: "As for me, I rather envy them. A world where it is always beautiful, where there are neither tempests nor cyclones, where the years are twice as long as ours, . . . where, in a word, everything is lighter, more delicate and more refined."

Percival Lowell gazes through his telescope.

THE MEN WHO CHASED MARTIANS

Flammarion's enthusiasm was shared by Percival Lowell, a budding American astronomer. Lowell came from a wealthy Bostonian family that had earned its fortune in textiles. (Lowell, Massachusetts, is named after Percival's uncle;[4] they were *that* kind of rich.) Textiles, though, weren't Percival's thing. He had loved astronomy for as long as he could remember, ever since witnessing Donati's Comet in 1858, when he was just three years old. He got his first telescope as a teenager, and his fascination with the stars continued into his college years at Harvard, where he studied mathematics but wrote his senior thesis on the formation of the solar system. In the years that followed, Lowell dutifully worked in his family's textile business before taking off for Asia, where he spent about a decade traveling and immersing himself in the Japanese and Korean cultures.

By the time Lowell returned to the United States in 1893, news had spread that Schiaparelli's eyesight was failing. The man who had witnessed the *canali* on Mars would no longer be able to see them. So, here was Lowell, back home and facing the high expectations that came with the family name. If he wasn't going to lead the business, how would he make his mark? Finding Martians seemed like a good plan.

Lowell merged his passion for astronomy with the sightings on Mars and his immense wealth and decided to open an observatory at a high elevation out west, where the skies were clear and free from city lights and smog. From there, he'd carry on Schiaparelli's work by mapping

4 The family's mills remain in Lowell, MA. So do the canals that were used to transport materials. One way or another, canals were integral to the Lowells.

the canals and discovering the intelligent people who built them. Like Flammarion, Lowell believed Mars was an older planet and was therefore further along in its evolution. The idea of canals aligned with this thinking and led Lowell to postulate that the Martians were trying to irrigate their dying planet by directing water from the polar icecaps.

With the 1894 Mars opposition fast approaching, Lowell sent a scout to Arizona and quickly settled on a 7,200-foot mountain peak in Flagstaff as the perfect spot to spy on Mars. Lowell named it Mars Hill. A railway running through town made it easy to have materials and equipment delivered. Construction began right away, but in order to start seeing the canals as they neared Earth, Lowell had to borrow an 18-inch telescope from Harvard's observatory.[5] There at his new home atop Mars Hill, he meticulously mapped out 184 waterways and continued devoting his time to studying the Red Planet and the Martians' digging frenzy (*below*). Just as importantly, he vigorously promoted his ideas through lectures and the press. By 1895, he had compiled an initial batch of studies and wrote the first of his three books on the planet, simply titled *Mars*.

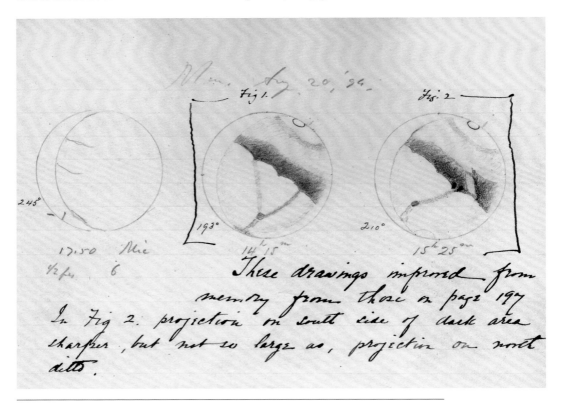

5 Lowell's custom-built 24-inch Clark telescope was built in time for the 1896 opposition. Visitors to the Lowell Observatory can still gaze through it every night.

IS MARS SIGNALLING?

The Startling Phenomena Seen by Prof. Percival Lowell and their Significance.

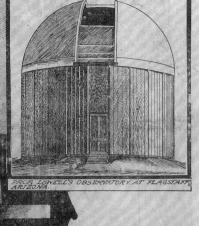

PROF. LOWELL'S OBSERVATORY AT FLAGSTAFF, ARIZONA.

THE "PROJECTION" OR SIGNAL FROM MARS, DRAWN FROM DESCRIPTION OF ONE OF THE ASTRONOMERS WHO SAW IT.

LAND

THREE SIGNALS SEEN THROUGH LICK TELESCOPE IN 1892.

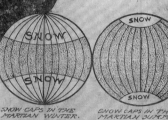

SNOW CAPS IN THE MARTIAN WINTER. SNOW CAPS IN THE MARTIAN SUMMER.

LATEST MAP OF THE CANALS OF MARS, DRAWN BY SCHIAPARELLI.

Marvelous "Projection" on the Planet, Discovered by Astronomers at Flagstaff, May Be an Attempt to Communicate With the Earth.

IS IT possible that the people on the planet Mars are endeavoring to make signals which will attract attention on the earth and pave the way to communication between these two orbs?

Is it possible that the strange, mysterious happenings on Mars, now reported by the greatest astronomer in the United States are, in fact, what he long has been searching for—artificial disturbances on a gigantic scale created by the men of Mars for the purpose of being seen on the earth, where they civilized race exists?

These are questions exciting discussion in scientific circles during the past few days since the astonishing information communicated by Prof. Percival Lowell of to the astronomical world.

FLAGSTAFF, Ariz., June 5.

LOWELL has been spending some months at his observatory making a prolonged study of of Mars, and is now authority for the statement that strange happenings observed by him on the planet are observed by him on the planet so confined almost all of his attention to the one planet. His object in this place during that time has gather given over to the study of

A LOCAL ASTRONOMER'S VIEW

MR. JOHN J. LICHTER, formerly lecturer on astronomy at Washington University, expressed much interest in the discovery by Prof. Lowell of a "projection" on Mars.

"I do not know just what to think of it," said Mr. Lichter, "as I am not in a position to express a scientific opinion just now. I have studied Mars to some extent, but have made nothing like the thorough study made for years by Prof. Lowell has given the subject. His assertions undoubtedly carry great weight.

"As to whether it be possible for the people of Mars—granting, for the sake of argument, that the planet is inhabited—to signal to the earth, astronomers are divided. There are some high authorities who admit the feasibility of such signalling.

"The revelations of Prof. Lowell are full of thrilling interest. We should not be quick to pronounce against the value of new discoveries. It is remembered that a few years ago we thought it would be impossible to photograph through the human body, but now almost any physician does it with the Roentgen ray. To science nothing seems impossible.

"I might say, however, that if the Martians indeed have signaled to the earth, by a great light or otherwise, the magnitude of the work which encompasses the result must be so enormous that it is difficult for earth-dwellers to conceive of it. Nevertheless, as I have said, we should not say a thing is impossible until we know it to be so."

TWO BROTHERS STOLE A POSTOFFICE

Case of Longwood, O. T., the Only One of Its Kind in the Records of the Postoffice Department.

GUTHRIE, O. T., June 4.

ALTHOUGH seemingly young and inexperienced the Burnett brothers, G. W. and R. F. were shrewd enough to fool Uncle Sam's postal inspectors for several months before it was ascertained that they had illegally obtained possession of an entire postoffice at Longwood, O. T.

Within the records of the postal department this is the only case of the kind, and it has caused widespread interest.

Beyond maps and data, Lowell filled his pages with logical explanations of how and why life should exist—with life on Earth offering his best arguments:

> *"When it had thus been conclusively proved that no life could exist at the bottom of the sea, deep-sea dredges were invented, and no sooner were they let down than, behold! they came up teeming with life. Fish and crustacea, mollusks and echinoderms—life, in short, of all the usual pelagic kinds from protoplasmic molecules to marine monsters—were found to inhabit the abyssal depths. What could not be, just was."*

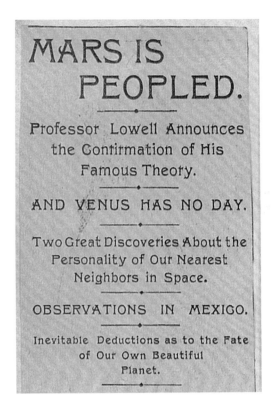

Why should Mars be any different? To Lowell, the logic and evidence was perfectly clear. Martians would've adapted to their conditions and evolved, just like us. Except, they'd surely look nothing like us.

So, if these Martians were evolved and superior in intelligence, why hadn't they reached out to Earthlings? If Flammarion was right, and they had sent messages during Earth's mammoth days, weren't they due for another try? Or were they too busy shoveling dirt and making red salads to bother us with some friendly conversation? Not according to Francis Galton. In 1896, the English statistician, sociologist, anthropologist, geographer, meteorologist, and author of more than 340 books and papers—not to mention Charles Darwin's cousin—thought the Martians were signaling Earth with a series of coded light messages. Someone just had to crack the code,

and he was determined to be the one to do it.

"That this is the means which people of Mars have finally adopted for opening communication with us there appears now to be no reasonable doubt, according to the statements of Sir Francis Galton," reported the *New York Journal* on January 17, 1897.

The crude lights were first detected two years earlier at the Lick Observatory in California on two successive nights. The signals returned, however, and remained active throughout the summer of 1896.

"The Martian telegraphers seem to have got the hang of their instrument—which must be built upon a gigantic scale—and to have devised a successful method of turning on and off the light in the immense area over which it must extend," the *Journal* wrote.

Galton may have been expecting such an event. Just four years earlier, in a letter to the editor of the *Times* in London, the curious scientist had suggested the notion of reflecting sun-signals to the Red Planet. "The inhabitants of Mars, if they have eyes, wits, and fairly good telescopes, would speculate on and wish to answer. One, two, three, might be slowly flashed over and over again from us to them, and possibly in some years, to allow time for speculation in Mars to bear practical fruit, one, two, three, might come back in a response."

Now that it appeared Martians had indeed adopted Galton's own plan for communication, he was determined to understand it. The hopeful codebreaker designed an apparatus to help him study the signals as he sat patiently next to the telescope. The machine used a long strip of telegraph paper which was slowly fed under a hinged pencil. Whenever Galton detected a flash of light he placed his finger on the pencil for the duration of the signal, marking the paper with a series of dots, dashes, and lines—a Martian Morse code, if you will.

Initially the markings were meaningless to him, but as the weeks passed, he believed he discovered order and regularity in the code which defined three distinct signals. Galton's diligent notes recorded thousands of impressions, observing that the three types of flashes lasted 1¼ seconds, 2½ seconds, and 5 seconds. Groups of them appeared to be separated by three seconds, indicating different words, whereas six-second separations designated the start of a new sentence. He documented all of his findings in an article entitled "Intelligible Signals between Neighboring Stars," published in the *Fortnightly Review,* one of the most prominent magazines in nineteenth-century England.

Galton determined from his measurements that the Martians had a numerical system based on 8, versus the base-10 method of counting. He had no explanation for the system of eights, but he did offer a theory given by one of his trusted assistants:

> *"A clever little girl who has helped us much by her quick guesses, entreats me to add her own peculiar view, which is that the Mars-folk are nothing more than highly developed ants, who count up to eight by their six limbs and two antennae, as our forefathers counted up to ten on their fingers."*

Her theory seemed as likely as any other.

But Galton moved on and continued to thoroughly dissect his dots, dashes, and lines, which led him to another theory. He found that the signals could be connected, with the outlines creating a form of picture writing. From this, he claimed to have translated twenty-nine words, including: area, brackets, circle, circumference, divided by, dodecagon, Earth, equal to, hexagon, Jupiter, Mars, minus, multiplied into, octagon, pentagon, perimeter of, picture-formula, radius, Saturn, square, sun, twelve-sided regular polygon, and triangle.

"Hello" did not appear to be in their vocabulary. The Martians, it seemed, were too eager to discuss math and planetary science.

But the subject of their conversation wasn't what excited Galton. It was the notion that a system of hieroglyphs could be understood by "inhabitants of neighboring stars in intelligible communication if they were both as far advanced in science and arts as the civilized nations of the Earth at the present time."

So why the sudden attempts to reach us? According to his *Fortnightly* article, Galton surmised that the Martian astronomers may have gotten help from "the money of a mad millionaire, or rather a mad billionaire" to develop their signaling device. Despite Galton's perceived accomplishments, no small talk about Martian weather or distinguished intergalactic greetings followed. More recent science has suggested these kinds of light flashes came from sunlight reflecting off water crystals in the Martian atmosphere while the planets were aligned. But in the late nineteenth century, people understood that long-distance relationships took time and effort, and considering the two planets had been around for billions of years, scientists were perfectly happy to keep at it for a few more.

SHALOM
FROM MARS!

Astronomers who wanted to see the canals could see the canals. After all, belief was far more powerful than any telescope. You could believe even more if you looked a lot closer. That's what happened in 1895 when a Washington scientist studying Percival Lowell's maps reported that an intricate arrangement of canals spelled out the Hebrew name of God. He identified the Hebrew letters shin, dalet, and yud, which reads Shaddai, a name used in Genesis when the Almighty speaks with Abraham.

The big revelation occurred while the scientist was reading one of Lowell's lectures, which reminded him of the recent light flashes from Mars. If Martians had tried to signal Earth with lights, wouldn't they try other methods, too? Like digging Hebrew letters into the planet's surface? That's when he pulled out the maps.

"The suggestion of some letter or cipher had hardly occurred when with it came the answer," reported one newspaper. "There it stood in bold letters, traced beyond a question by intelligent beings as a greeting and an overture to the people of our earth for mutual study and friendship."

This astute observer was no religious fanatic. Rather, the paper assured its readers, "He was a frank agnostic and his observation was therefore unbiased by any religious goal."

Okay, but how could beings carve such a message?

"True, the magnitude of the work of cutting the canals into the shape of the name of God is at first thought appalling," the paper acknowledged, "but there are terrestrial works which to us to-day seem no less impossible." Plus, if anyone could do it, it was those masterful Martians astronomers had been raving about. That didn't explain how they knew Hebrew though. "Perhaps [the scientist] suspects they may be the descendants of the lost tribe," another reporter suggested.

The whole notion of seeing God's name on Mars may sound absurd, but the idea of seeing what you want to see is human nature, and that certainly hasn't changed. People do the same thing today while scouring NASA images of Mars. The Sphinx, human faces, and various fossils have all been spotted. Mars, it seems, is as human as we make it. More on that in the next chapter.

French amateur astronomer, A. Mercier, thought the Eiffel Tower might be just what was needed to create a successful signal. In a booklet called "Communications with Mars," he suggested attaching several large reflectors on the structure, aiming them at the planet. A large screen could interrupt the beam of light to indicate a "non-natural" message.

Martian hoopla and enthusiasm had taken the world by storm, but some weren't so quick to jump on the bandwagon. A review of Mercier's text questioned his evidence about intelligent life and suggested that, if it did indeed exist, "we may just as well assume that the problematic dwellers on Mars are reptiles, occupied in eating each other up."

Of course, who's to say such cannibalistic reptiles weren't intelligent, but belligerent beings? Whatever they were, people wanted to believe in Martians and our ability to contact them. After all, technology was moving fast and the world was changing. Alexander Graham Bell had just simplified communication with the telephone, and demand for his machine swelled wherever people wanted to talk to other people. So Bell looked to expand, naturally, to Mars. In the late 1890s he began working on a new invention called the "photophone." Sound, Bell believed, could be carried through rays of light, meaning he could transfer messages endlessly across the vacuum of space. The inventor felt confident in the principles of his design, but despite all the scientific genius on Earth, the photophone, just like the telephone, required someone on the other end to receive the call. Before Martians could answer, they needed to know how to build a receiver.

That didn't discourage Bell, though. Reporters remained optimistic, too, including one who proclaimed, "The theory is flawless." With all the recent advances, why not think big and reach further? As one scientist put it in 1901:

"The general reading public, in its insatiable craving for highly-seasoned pabulum, is, of course, most willing to accept the explanations of the [Mars] enthusiasts. With telegraphs, railways and fast steamers to the remotest parts of the Earth, and special correspondents and camera-fiends even in the most God-forsaken spots of No-Man's Land, the globe is fast becoming too small, and the daily budget of news so monotonous, that nothing but copy from other worlds can save the coming generation from being bored to death."

If the general reading public liked what papers had been feeding them about Mars, they must've loved what *Collier's Weekly* served up in a 1901 issue. Nikola Tesla explained that he could contact Martians by accurately sending wireless messages to any point on the Red Planet. Yes, Tesla. The guy who made it possible for you to plug in stuff at your house, like your phone or maybe your car, thanks to the concept of alternating current (the AC in AC/DC). The electrical engineer received three hundred patents for his inventions. Think what you will about Charles Cros and Camille Flammarion, but Tesla was no slouch.

Just two years earlier, while conducting experiments with his Magnifying Trans-

NICOLA TESLA PROMISES COMMUNICATION WITH MARS.

NICOLA TESLA

mitter in the high altitude of Pike's Peak in Colorado Springs, Tesla claimed to have detected signals that originated on Mars. "Inexplicable, faint, and uncertain though they were," he was certain this was the moment humanity had been waiting for. "Brethren we have a message from another world, unknown and remote. It reads: ONE—TWO—THREE."

Tesla's article in *Collier's* gave no specific thoughts on what kinds of beings the Martians might be, only that he believed creatures could adapt to the planet's conditions and thrive. "I think it quite possible that in a frozen planet, such as our moon is supposed to be, intelligent beings may still dwell, in its interior, if not on its surface," he said.

Other scientists, though open to the possibilities of Martian life and the ability to one day enjoy interplanetary communication, were skeptical of these claims. "I believe that Tesla is an exceptionally brilliant man," said Professor A. N. Skinner of the Naval Observatory. "He has done much wonderful work, but I take no stock in his new proposition, and, indeed, I have talked with no scientific men who do not consider Tesla's claim absurd."

Yet Tesla's conviction, along with the beliefs of Lowell, Flammarion, and other contemporaries, convinced newspapers that their thoughts were as good as facts. Skinner's pessimism wasn't exactly great for sales, so papers had no qualms about declaring that "human life on the planet Mars is established beyond the shadow of a doubt."

If Tesla was right, and Martians were trying to send messages, how might we receive them and in what other ways might we try to communicate with them? By 1904, the engineer said he had the answer but remained elusive about the details. "I made the discovery a year and a half ago," Tesla told reporters. "Although I am ready to talk with the people of Mars, I shall not tell how soon the talking shall begin. All will be told later."

PROFESSOR WIGGINS AND THE
MARTIAN METEOR TOSSERS

The belief in Martian canals or perceived wireless signals gave many scientists the proof they needed to claim that the Red Planet was inhabited. But Ezekiel Stone Wiggins looked to a different source: meteorites.

Wiggins was a Canadian amateur astronomer and a weather prognosticator, known for predicting massive storms. His work earned him the name "The Ottawa Prophet," although few of his predictions ever came to pass.

In November of 1897, after a meteorite landed in Binghamton, New York, Wiggins theorized that it had been hurled from Mars and contained a hieroglyphic message engraved on its surface. In fact, he was convinced that Martians had been tossing rocks to Earth for quite a long time.

"My opinion is that stones have for many thousands of years fallen from space upon the Earth which actually contained written characters," he told reporters. "The ancient Jews and other nations speak of the sacred books as having fallen from heaven. As the earliest important records were preserved in stone, it seems probable that the idea originated with aerolites like that of Binghamton. There is no doubt in my mind that there are thousands of these stones that have fallen to our planet since man arrived here and are messages from another planet."

To bolster his theory, Wiggins shared tales of other odd occurrences throughout history, including stone showers mentioned in the Bible (striking the enemies of Joshua) as well as in Rome in 652 BCE and again in 705 BCE. He also discussed a 54-pound meteorite that hit France in 1637. "It had the size and shape of a human head," he said. Even more curious was the black stone housed in the British Museum that "has a portrait of the poet Chaucer."

Given the hype about the canals and Francis Galton's study of Martian light flashes, Wiggins felt justified in his thinking. "Why should not the Martians throw missiles to the Earth to inform us of their existence and condition, when they form enormous lights and geometrical figures on the surface of their planet to attract our attention?"

So, how exactly did Martians manage to throw rocks millions of miles? Wiggins first assumed they had a greater intelligence than Earthlings do, which simplified the feat. In a few hundred years, he thought, we too might have enough smarts to do the same thing.

"Suppose the Martians were to throw a stone highly electrified into the orbit of their nearest satellite, which is only about 7,000 miles away, so that it would be in advance of it in its orbital motion," Wiggins explained. "I have no doubt it would repel the stone in the line of a tangent and with such force as to send it to our planet's orbit, or, suppose a comet were passing near Mars and toward the Earth, stones thrown near it would follow in its trail and fall to the Earth like the stone which fell to the Earth in November, 1872, after the comet of that year had crossed our planet's orbit."

A year later, one journalist reporting on the astronomer's theory said that if Wiggins was right, "the Martians are an ill-bred race, and it is not worth our while to know them."

While we humans waited for Tesla's big plans, other hopeful thinkers began to brainstorm answers. In 1909, they were ready to be put to the test. Mars was in opposition and would be closest to Earth—just 36 million miles away—in fifteen years. Perhaps expanding on Cros's idea, Harvard professor William Henry Pickering considered signaling the planet with a system of reflections made by fifty giant mirrors.

"As far as the people of Mars are concerned this reflector would not, of course, be apparent to the naked eye, but through lenses of such magnitude as we have today the reflection would be easily discernible and would undoubtedly attract attention at once," Pickering explained.

He believed the signals would need to continue for three to four months and be repeated for several years. This, the professor felt, should give intelligent life plenty of time to notice the messages and construct a simi-lar means of sending one in return.

The assembly and operation of such a grand plan, however, was estimated to cost a cool $10 million. Today, that would be equivalent to more than a quarter billion dollars.

The press offered a suggestion: Convince philanthropist Andrew Carnegie to supply the necessary money. Carnegie, one of the richest men in America, gave away nearly $350 million to various universities, charities, and foundations in the last eighteen years of his life. So what's another ten mil?

Pickering had, in fact, gone to Carnegie seven years earlier to make a pitch for a Carnegie Astronomical Fund in connection with the Harvard Observatory. The wealthy industrialist said no.

"At this point Professor Pickering would find it necessary to persuade the great capitalist who is striving to die poor that the $10 million would not be flashed off into space and never be heard from again," a newspaper wrote. "In short, the professor must demonstrate a reasonable probability that the Martians are on the watch waiting for us to open up communication with their own planet, and would understand the Earth's signals."

If the Martians were as advanced as many thought, success could bring knowledge that might advance our own species by thousands of years.

"It would be enormously helpful to hear how Mars settled his race questions and composed the quarrels of capital and labor. . . . One can think of a thousand questions to flash up to the Martians—how they perfected flying machines, how they cured cancer, what finally became of war, whether they have any use for lawyers. . . . If Mr. Carnegie has any imagination, how can he cast aside such a stupendous opportunity?"

Apparently, Carnegie's imagination was nonexistent when it came to Mars. Or perhaps he was dissuaded by Tesla's a letter to the *New York Times* rejecting Pickering's idea. In short, the mirrors would never generate enough power to reach the Martians. Tesla did, however, reiterate his beliefs that Mars was reachable through a wireless transmitter.

"In my experiments in 1899 and 1900, I have already produced disturbances on Mars incomparably more powerful than could be attained by any light reflectors, however large," he wrote.

Professor Harold Jacoby, head of the astronomy department at Columbia University, sided with Tesla. He told reporters that Pickering's idea was "scientifically possible, but impractical." Jacoby recommended we wait for Martians, if any existed, to contact us first. "Would it not then be time enough to rig up an answering apparatus?"

Jacoby's associate, Dr. Samuel Alfred Mitchell, agreed. The two seemed to think the whole thing was silly but appreciated the interest in astronomy created by the conversation. Adding fuel to the fire, Mitchell offered his description of what a Martian might look like: "To begin with, he would be tall and spindle-shanked. It is a certainty that there are no fat men on Mars. The attraction of gravity is two-thirds less there than here, hence he would grow upward instead of sideways. He would make a great marathon runner, as the resistance is less. His head would be immense when compared to the diameter of his body and his eyes might be as big as saucers. . . . I don't believe the men are web-footed, but they probably grow a fine crop of fur."

Pickering never got his $10 million, but one of his peers offered a cheaper alternative. Professor Robert Wood of Johns Hopkins University believed a series of giant black spots across the open space of

Nevada could be used to make Earth "wink" at Mars:

> *"A large black spot on the white alkali plains could be constructed at much less expense, and would be as easily perceived by the Martians, if they exist and have telescopes as powerful as ours. It would be as easy to 'wink' signals with the black spot as with a mirror of equal size, probably easier. The spot could be made in small sections of black cloth arranged to roll up on long cylinders, exposing the white ground underneath, the cylinders being operated simultaneously by electric motors."*

A little interplanetary flirting with Martians gazing through telescopes might be just the thing to break the ice. Maybe they'd construct a similar "come hither" system, or at least give us a thumbs up (if Martians had thumbs).

All Wood needed was a little funding for a lot of fabric. "I am unable to say how much four square miles of cloth would cost," he added. "You will have to consult the dry goods houses or the people who write arithmetic."

Despite the bargain, Earth never winked.

SEVEN THINGS TO ASK MARTIANS IN 1909

When Pickering's plan to contact Mars went public, the public responded. *Harper's Weekly* offered some questions to ask once his "flashophone" found success, ensuring that any opportunities for conversation wouldn't go to waste:

1. Is it hot enough for you?
2. Have you seen the man higher up lurking around any of your tow-paths?
3. Should women smoke, and if so, what?
4. From where you are does the Earth look like a pinhole in a rubber blanket, or a microbe floating around through space?
5. Have you planted your early spring beets yet?
6. Are your canals locked or on the level?
7. Whom do you consider the greatest living American?

IS THERE LIFE
ON EARTH?

In 1906 Professor Edward S. Morse, director of the Peabody Academy of Science, spent thirty-four nights at the Lowell Observatory staring at Mars, which he then wrote about for the October 7 issue of *The World Magazine* (*right*). So, after more than a month, did he believe the Red Planet was inhabited? His conclusion, in a word: "Unquestionably."

Like Percival Lowell, Morse countered arguments about life being unsustainable on a planet with so little atmosphere by reminding people of the extreme conditions in which life thrives on Earth. To Martians, those strange and extreme conditions would simply be normal. In fact, they might even be looking at Earth wondering how we could possibly live in conditions that might seem uninhabitable to them.

To illustrate his point, Morse offered these thoughts:

What the Martian Astronomer Might Say of the Earth

- Many facts which have come under our observation tend to support the belief that no life can possibly exist on the earth.
- The tremendous force of gravitation, with great atmospheric pressure, would forbid the existence of any organic forms.
- The immense clouds veiling the surface must at times suffer condensation and the impact of raindrops would from their velocity and weight smash everything in the way of life.
- Life if it existed in forms supported by appendages must have legs of iron to sustain its weight, and a crust like a turtle's to be impervious to raindrops.

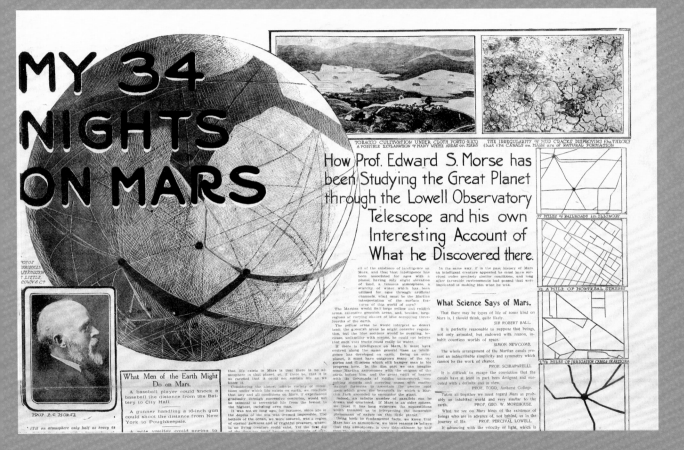

MY 34 NIGHTS ON MARS

How Prof. Edward S. Morse has been Studying the Great Planet through the Lowell Observatory Telescope and his own Interesting Account of What he Discovered there.

TOBACCO CULTIVATION UNDER CLOTH, PORTO RICO
A POSSIBLE EXPLANATION OF MANY WHITE AREAS ON MARS

THE IRREGULARITY OF MUD CRACKS DISPROVING the THEORY that the CANALS on MARS are of NATURAL FORMATION

37 MILES OF RAILROADS in ILLINOIS

½ A MILE OF MONTREAL STREETS

1¾ MILES OF IRRIGATION CANALS IN ARIZONA

PROF. E. S. MORSE

What Men of the Earth Might Do on Mars.

A baseball player could knock a baseball the distance from the Battery to City Hall.

A gunner handling a 16-inch gun could shoot the distance from New York to Poughkeepsie.

A pole vaulter could spring to

ITH an atmosphere only half as heavy as

ell of the existence of intelligence on Mars, and that that intelligence has been associated for ages with a planet having only slight elevation of land, a tenuous atmosphere, a scarcity of water which has been utilized for ages through artificial channels, what must be the Martian interpretation of the surface features of this world of ours?

The Martian would find large yellow and reddish areas, extensive greenish areas, and, beside, large regions of varying shades of blue occupying three-fourths of the earth.

The yellow areas he would interpret as desert land; the greenish areas he might consider vegetation, but the blue sections would be puzzling, because, unfamiliar with oceans, he could not believe that such vast tracts could really be water.

If there is intelligence on Mars, it must have evolved along the same general lines as intelligence has developed on earth. Being an older planet, it must have outgrown many of the vagaries and illusions which still hamper man in his progress here. In the dim past we can imagine some Martian astronomer with the enigma of the earth before him and the great vault of heaven with its thousands of riddles unanswered, computing records and covering pages with mathematical formulas to ascertain the precise spot upon which grew the beanstalk by which a Martian Jack ascended to encounter the giant.

Indeed, an infinite number of parallels can be drawn and quartered. If Mars is an older sphere, we trust it has long outgrown the superstitions which trammel us in interpreting the inexorable phenomenon of nature on this little planet.

Going back to fundamental facts, we know that Mars has an atmosphere; we have reasons to believe that this atmosphere is very thin—thinner by half than the atmosphere

that life exists in Mars is that there is no atmosphere in that planet, or, if there be, that it is so rarefied that it could not sustain life as we know it.

Considering the almost infinite variety of conditions under which life exists on earth, we conclude that any and all conditions on Mars, if experienced gradually through successive centuries, would not be inimical to terrestrial life from the lowest to the highest, including even man.

It was not so long ago, for instance, since life in the depths of the sea was deemed impossible. The bottom of the ocean, we were assured, was a region of eternal darkness and of frightful pressure, where-in no living creature could exist. Yet the first dip of the deep sea trawl brought up animals of all

In the same way, if in the past history of Mars an intelligent creature appeared he must have survived under precisely similar conditions, and long after favorable environments had passed that were implicated in making him what he was.

What Science Says of Mars.

That there may be types of some kind of life on Mars is, I should think, quite likely.
SIR ROBERT BALL.

It is perfectly reasonable to suppose that beings, not only animated, but endowed with reason, inhabit countless worlds of space.
SIMON NEWCOMB.

The whole arrangement of the Martian canals present an indescribable simplicity and symmetry which cannot be the work of chance.
PROF. SCHIAPARELLI.

It is difficult to escape the conviction that the canals have at least in part been designed and executed with a definite end in view.
PROF. TODD, Amherst College.

Taken all together we must regard Mars as probably an inhabited world and very similar to the earth.
PROF. GEO. W. MOREHOUSE.

What we see on Mars hints of the existence of beings who are in advance of, not behind, us in the journey of life.
PROF. PERCIVAL LOWELL.

If advancing with the velocity of light, which is

A colleague at Amherst College, Professor David Todd, took a much different approach. Rather than construct giant contraptions on the ground, he planned to set up shop high in the sky—at 50,000 feet, to be specific. By developing a hot-air balloon that could soar to such heights, he believed signals could be received from Mars without interference.

"We shall shut ourselves in an aluminum cage fitted with apparatus to drive out carbonic acid gas and supply oxygen and air pressure to prevent mountain sickness," Todd explained to reporters, regarding his balloon plans. "We shall ascend as high and stay as long as possible. With our wireless apparatus we shall seek not to send but to receive messages from Mars and Venus." (Yes, Todd believed that Venusians, too, had been trying for years to communicate with Earth.)

The professor further explained his plan for achieving the ground connection required for telegraphy: "Perhaps a thousand-foot wire hanging from the car would be sufficient, with the surrounding atmosphere, to form a grounding. If not we shall use ten miles of piano wire and allow it actually to touch the ground."

Success would not only benefit science, but it would also make us mere humans look good in the eyes of our alien neighbors. Conversely, if his venture failed, he'd at least achieve the longest tether in the history of, well, tethering.

"If life really exists on Mars, they have been trying for years to get into conversation with us, and perhaps wonder what manner of stupid things we are not to respond," Todd said in May 1909, months before his proposed September trip. "If the Martians are sending wireless [messages] I'll receive them; if they are not, I'll simply have spent $1,000 on a very interesting if rather hopeless experiment."

By August of that year Todd and his wife took their first balloon test flight. They reached a height of 5,000 feet. Though they carried no Martian-contacting equipment with them, they deemed the trip a grand success and Todd gained confidence in his plan. Sadly, his big day never arrived. September passed without a magical balloon ride to the Martian heavens.

It was yet another plan that went nowhere as 1909 came to an end without a peep from the Martians. But as far as Percival Lowell was concerned, they were just preoccupied. In 1910, newspapers reported Lowell's announcement of "the completion of a new gigantic engineering enterprise by the people of Mars, who are 'making the dirt fly' in a manner that excites the envy of the builders of the Panama . . . By his count, there were 690 canals.

The impassioned astronomer photographed the Martians' work and measured the waterway to be a thousand miles long. "Water has been turned—into it, and between the months of May and September vegetation has appeared in an hitherto uninhabitable part of the great desert which spreads over the greater part of the planet's surface," a report claimed.

Lowell had taken photographs the previous year, which did not show these canals. Therefore, he was convinced the phenomenon couldn't be natural and had to have been created artificially.

"The canals in Mars can be seen by children," he said. "They were not there in June; they were there in September. They are still there."

Professor W. S. Burnham of Wisconsin's Yerkes Observatory agreed with Lowell's theories. So did many journalists, one of whom wrote, "Professor Lowell's latest discovery is of the utmost importance and supplies the strongest confirmation yet obtained of the theory

CAN MARS SOLVE WORLD MYSTERIES

Daring 10-Mile Ascension Through Earth's Air Cushion May Answer—Professor and Aeronaut to Make Trip

AMHERST, Mass., June 14.—A cushion of air, 10 miles thick, surrounds the earth.

Out there, beyond this air cushion, there may be a veritable clamor of messages from Mars. For centuries these messages have been beating against this cushion; they have been beating in vain; they are hurled back into ethereal echoes that are wasted on space.

In vain the Martians may have strengthened their wireless message machines. With each strong-

gin to use the tanks at a height of about five miles," says Stevens, who has held long conferences with Prof. Todd, at Amherst. "I don't know much about astronomy, but I have agreed to take Prof. Todd where he wants to go, in my car.

"I think we can reach the height we seek in about six hours. The whole wireless apparatus, including the wire, will weigh only 200 pounds.

"Of course the trip will be dangerous. The greatest known

PROF. DAVID TODD AND WIFE. LEO STEVENS, AERONAUT. DIAGRAM SHOWING HOW BALLOON MAY ASCEND ABOVE AIR CUSHION, AND HOW WIRES WILL CARRY MESSAGES TO EARTH.

that there are intelligent beings on Mars."

As fun and exciting as all this talk about canals was, there were those who had heard enough of Lowell and his supporters. English scientist Alfred Russell Wallace thought the whole thing was absurd on account of simple physics. Any water being irrigated would've evaporated before reaching its destination.

"The mere attempt to use open canals for irrigation purposes would argue ignorance and stupidity," he claimed. "Long before half of them were completed their failure to be of any use would have led any rational being to cease constructing them."

Wallace made a good point, but Lowell didn't care. He wasn't about to let another scientist ruin his fantasy and his life's work. Nor would he entertain arguments from others who proposed there never were any canals to begin with. Like English astronomer Edward Walter Maunder, who proposed the lines on Mars were nothing more than an optical illusion. To prove it, he drew circles and marked them up with irregular dots, then placed them far away from a group of children and asked them to draw what they saw. The children drew circles with lines, or "canals," just as Lowell had done. It seemed Lowell was right about at least one thing: even kids could see the canals.

Eugene Antoniadi, director of the Section for the Observations of Mars of the British Astronomical Association, had once been a believer

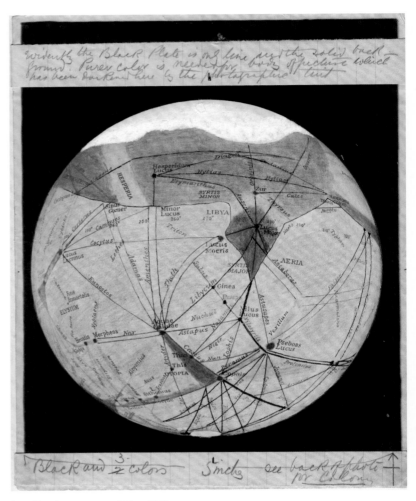

Lowell's notes on a map of Mars, 1905.

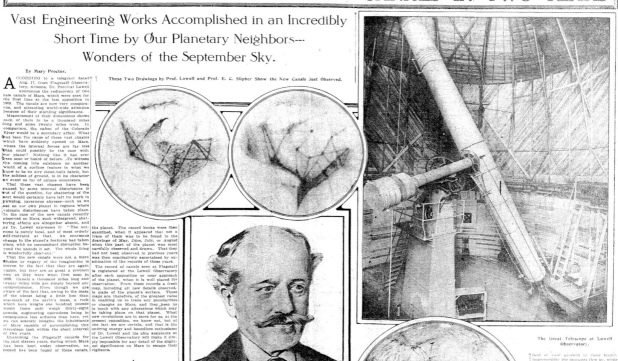

but now agreed with Maunder. A good look at Mars through Europe's most powerful telescope during the opposition also helped change his mind. Antoniadi issued a lengthy report stating that, in a nutshell, Mars had no canals. Those lines everyone was raving about? "Merely a summation of complex details," he wrote.

Lowell had always brushed aside such criticisms. In *Mars as the Abode of Life*, he defended his observations and mapping of the canals, adding, "I say this after having had twelve years' experience in the subject—almost entitling one to an opinion equal to that of critics who have had none at all."

"Fancy a company of Martian laborers, imported from their distant planet to dig the Panama Canal. ... Digging such a little ditch would be a matter of merely a few weeks' exercise."

—*Washington Times*, 1907

France had begun work on the Panama Canal in 1881, but the construction was taken over by the United States in 1904; the canal opened in 1914.

IRRIGATING AN ARID WORLD

LIFE SUSTAINED ON MARS THROUGH THE MEDIUM OF ITS WONDERFUL CANALS

Camille Flammarion

Surface of Mars Showing the Lines of the Canals.

Prof. W.H. Pickering.

Prof. Percival Lowell.

FANCY a company of Martian laborers, imported f r o m their distant planet to dig the Panama Canal. How the dirt would fly!

Digging such a little ditch would be a matter of merely a few weeks' exercise. Each Martian, according to the most recent estimates of scientists, could toss over his shoulders two and a half tons of a saucerful—and the supreme achievement of the race is canal digging.

For the giants of Mars are canal makers through stern necessity. It is not for purposes of commerce that they have lined and ribbed their dying planet with a vast system of waterways. Only through the most gigantic network of canals conceivable are they able to keep life in their arid world and provide sustenance for themselves.

Scientists now declare that the many lines and dark spot on Mars represent verdure along a most wonderful canal system, which the inhabitants of the planet have constructed for purposes of irrigation.

Through their artificial waterways the melting snow and ice of the poles are carried to various parts of the planet, and so the Martians are able to raise their crops in season and to stay, for a time, the menacing dry decay of their world.

S EVERAL of the greatest present-day astronomers—Lowell, Pickering, Flammarion, Morse and others—have practically agreed on the theory that Mars contains no rivers, lakes, oceans or any other source of water supply than the great fields of snow in the north and south poles.

At the summer time, when the people melt most proportion of snow or starve, the melting snow from these extremities of the sphere is carefully apportioned by means of artificial waterways to the "irrigation sections where it is needed. The Martians thus make the water run up hill, it is declared.

Professor Percival Lowell believes that practically the source of water supply for the planet is always polar. So long as the ice remains over the equator there is no water.

On Mars the vegetation spreads from the poles, because the snow must melt and the water flow into the canals before venture can grow. And in this connection Professor Lowell has made one of the most astounding of all his statements. He says.

"The quickening proceeds rapidly, and very nearly, if not quite, uniformly down the disk. It takes the darkening only five days to descend from the sixty-fifth parallel to the equator, a journey of 2,650 miles. This means a speed of fifty-three miles a day, or two and two-tenths miles an hour. And it does this in the face of gravity."

In fact, in plain language, the astronomer declares that not only does human intelligence in Mars direct the vast canals thousands of miles in length, but that, by some mysterious method, it causes the water to run up hill.

Professor Camille Flammarion, who has made a careful study of the planet, has been for several years watching the snowfalls on Mars. He believes that the Martian seasons may be subject to as many vicissitudes as ours, instead of being actuated by the exceptional constancy hitherto attributed to them. He agrees that the canal systems are artificial and were constructed with a view to irrigation.

What manner of people are those who do such remarkable things? Obviously, a quite different kind from the inhabitants of this earth.

According to the best authorities, founded on the most recent investigations, the Martian is a creature immensely more powerful, physically, than earth mortals, even earth giants. This is deduced from the lesser pull of gravity on Mars.

A Martian could run 100 yards in three or four seconds, could leap over a high tree, could kick a football a quarter of a mile.

Because of the lesser attraction of gravity he may be at least three or four times as big as the average human being, perhaps even much larger than that. As their rising wheels, perhaps, adds weight to the living in that, on account of the rarefied air on Mars, a Martian must require three times as much lung space as an earth mortal to get enough air to live, and his body must be proportioned accordingly.

TURNS OUT MARS IS A VEGETABLE WITH A GIANT EYEBALL WATCHING OVER IT

All the hype about canals and vegetation on Mars led to a new theory published in 1912 by the *Salt Lake Tribune*. It claimed that Professor William Wallace Campbell, director of California's Lick Observatory, believed that Mars was a vegetable. "This theory is one of the most plausible that has been put forward," the paper wrote.

You see, all those canals that scientists had been seeing were lines of yellow and orange vegetation, which gave Mars its reddish color. "The vegetation theory rests primarily on the fact proved by spectroscopic analysis that there is an enormous proportion of carbonic acid gas in the atmosphere of Mars," the story explained, "which would make animal life of the kind known to the Earth exceedingly difficult, if not impossible, while it would greatly favor the development of vegetation."

The theory reasons that just as carnivorous plants on Earth have animal-like abilities, the evolution of vegetation on Mars gave it a nervous system and intelligence. It eventually became "one organic whole in order to obtain the greatest advantage from the limited means of sustenance. Life on Mars is now one vast vegetable body having its roots in the soil."

But wait, there's more. The white patches on the Martian surface that astronomers thought were piles of snow weren't snow at all. Campbell's theory called it an "eye." This Martian peeper was "supported on a tenuous flexible transparent column" and could "raise itself miles above the surface of the planet and watch operations of its vegetable body at any point. When not engaged in watching the physical condition of its body, the great 'eye' makes observations of the Earth, sun, planets, stars, and the whole universe. From its vast side it is able to see more and farther than all the telescopes of our Earth put together."

The story was a hoax, and Campbell wasn't too happy about it; years earlier he'd said he didn't believe Mars had an atmosphere capable of supporting life. But despite the fabrication, the Tribune's article was a truthful demonstration of how Martian fanaticism was sweeping the Earth.

Magazine Section

The Salt Lake Tribune.

Sunday, October 13, 1912

Mars Peopled by One Vast Thinking Vegetable!

Interesting Theory of Prof. Campbell, of Lick Observatory, That Explains the "Canals," "Eyes," and Other Puzzling Problems of Our Neighbor Planet

ODD FACTS ABOUT MARS

MARS is the fourth planet from the sun, and the nearest to our earth.

It is called the red planet, and its color is thought to be due to vegetation.

Its size and density are less than ours, and a man weighing 200 pounds here would only weigh seventy-five pounds there.

Mars has atmosphere, seasons, land, water, storms, clouds and mountains.

Mars has two [illegible] only 3,700 miles away and revolves around it in seven and a half hours [illegible] shooting star.

The day on Mars is half an hour longer than ours, and its year contains 687 days.

Professor Lowell has counted 437 "canals" on Mars, and 186 "oases." The canals vary in length from 250 miles to 3,000 miles.

A man on Mars would be able to drive a golf ball fifty miles.

The strength of a man on Mars would be eighty-three times greater than on the earth.

The atmosphere of Mars consists principally of carbonic acid gas.

The water supply of Mars is very slender, and its utilization is the greatest problem of life there.

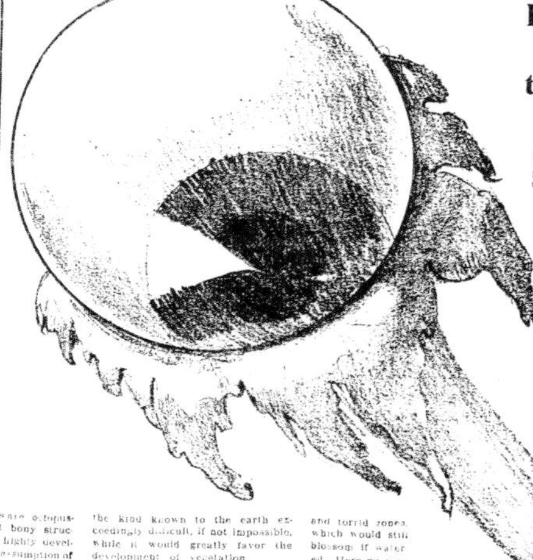

MARS is the nearest planet to us, the one which we can see the most plainly and the first one which we shall be able to visit when science makes a journey beyond our atmosphere possible.

Mars, moreover, is proved by astronomy to possess air atmosphere and to be capable of supporting life in some form.

These facts make it natural that scientists and writers should speculate upon the character of the life upon Mars. It must as yet be speculation, for our means of seeing only enable us to distinguish objects several miles in extent upon the planet.

Many interesting theories about the life of Mars have been put forward, but all of them have been open to some objection. Professor Percival Lowell, of the Flagstaff Observatory in Arizona, has argued that the so-called canals of Mars are vast engineering works, and consequently that the inhabitants who built them were of great intellectual development. The scientific novelist, H. G. Wells, has built an extremely interesting story on the basis that the Martians are octopus-like creatures, without bony structure, but possessed of highly developed brains. A common assumption of many speculators has been that the Martians are extremely attenuated creatures, because the slight pressure of gravity on the surface of the planet would favor this form.

Now a new and exceedingly interesting theory concerning the life on Mars has been put forward by Professor William Wallace Campbell, of the great Lick Observatory, California. He suggests that all life on Mars has taken a vegetable form.

This theory is one of the most plausible that has been put forward. It has the support of all the facts about Mars that have been scientifically established and it avoids many of the improbabilities involved in other theories on the same subject.

The vegetation theory rests primarily on the fact proved by spectroscopic analysis that there is an enormous proportion of carbonic acid gas in the atmosphere of Mars, which would make animal life of the kind known to the earth exceedingly difficult, if not impossible, while it would greatly favor the development of vegetation.

Before considering this theory further, we must bear in mind a few of the proved facts about Mars. It has atmosphere, seasons, land, water, storms, clouds and mountains. It also rains and snows on Mars, as it does with us. Great white patches appear periodically upon its surface. These may be accumulations of snow and they have also been called "eyes." Their nature will be discussed later.

When Mars approaches nearest to the earth it is seen to have a bright red color and sometimes looks like a red lamp in the sky. It has been suggested that the vegetation for the most part is yellow or orange instead of green, as with us, thus giving the planet its color.

Mars has two moons, the nearest of which is but 3,700 miles away and revolves around the planet in seven hours and a half, showing all the phases of our moon in one night.

The density and size of Mars are less than those of the earth and consequently a man who weighed 200 pounds here would only weigh 75 pounds up there.

The atmosphere and moisture of Mars are very slight, and the inhabitants, if there be any, must find life a difficult problem there. The water is confined entirely to the poles, where it is deposited annually in the form of a thin lawyer of snow or hoar frost, only to melt away again with the advent of Summer.

Ages ago life on Mars must have concentrated itself on the problem of devising some means whereby the melting water of the polar parts might be conducted to those arid regions of the temperate and torrid zones, which would still blossom if watered. Here we may note one of the strongest arguments in favor of the vegetation theory of life.

Professor Lowell has argued that the canals of Mars, first discovered by Professor Schiaparelli, of Milan, but long considered optical illusions by many astronomers, are the irrigation works of the inhabitants. The canals are singularly artificial in appearance. They extend toward the equator from the poles and cover the planet like a fine netting.

Each canal is the shortest distance between two points and invariably runs to a point called "an oasis," where it meets other canals not in haphazard fashion, but according to some plan.

The "canals" vary in length from 250 miles to over 3,000 miles, a length that is astonishing when it is considered that the diameter of Mars is only 4,229 miles. All told, Professor Lowell has plotted 437 of these canals and 186 oases. It was Professor Pickering, a close associate of Professor Lowell, who first saw these oases.

The canals of Mars appear and disappear with the seasons. In other words they slowly creep down from the poles each Spring and slowly retreat with the approach of Winter. If the canals are artificial in origin, this phenomenon would appear to mean that the Martians are busily engaged in digging stupendous canals only to fill them up again every year.

Professor Pickering ingeniously avoided this embarrassing conclusion by pointing out that we see not the canals themselves, but the vegetation which fringes their banks and thus indicates their course. Vegetation must grow before the canals are visible and must disappear before the canals vanish.

Here we can see why the theory that all the life on Mars is vegetation is more probable than any other. Professor Pickering admits that the signs of life we see are vegetable, but suggests that they are the work of man-like creatures whom we cannot see. But we know that it would be almost impossible for man-like creatures to live there. It is most likely then that vegetation is the only life.

With our knowledge of life on the earth, it is quite conceivable that the highest type of intelligence might dwell in a plant, as plants that we know possess more or less intelligence, and the fact that they may not possess the highest kind is due to conditions on the earth which do not exist on Mars.

The original germ of life on earth was neither animal nor vegetable. Many stages of development passed before the two forms of life became separated. Now in the scale of life we now see many forms of which it cannot be said positively that they are vegetable. When disease bacteria were first observed it was believed that they were animal, but now the prevailing view is that they are vegetable. Yet they possess the power of motion generally characteristic of animals.

One of the great differences between plants and animals is that the former have not a brain and nervous system, which can be compared to that of the latter. The life of the plant resides in its separate cells and they are only held together by their juxtaposition to one another and not controlled by a central system.

But there are some plants on earth which do possess a kind of nervous system, and it is quite reasonable to believe that they would have developed an intelligence at least equal to that of man if conditions had been favorable. Such conditions have prevailed on Mars. Chief among them is an atmosphere very favorable to plant life and very unfavorable to animal life.

There are on the earth many carnivorous plants which though unmistakably vegetable in form possess many of the powers of animals. These plants include the butcher plant, pitcher plant, the sundew, the butter wort and many other forms. They range from plants that eat insects to those that are capable of devouring birds and small mammals.

The pitcher plant, for instance, has a heavy flesh leaf ten inches long. With the spiked point of the leaf it strikes a rat, numbing it with the poison it contains. Then the leaf folds over the animal and it is absorbed into the body of the plant and digested.

Other plants, such as the spider-wort, possess eyes, which enable it to turn toward the sunlight. These eyes bear a close resemblance to human eyes and it has been proved by photography that they receive images of objects which lie in their range of vision.

These facts indicate the possibilities of vegetable intelligence. Ages ago, according to the newly-advanced theory, all life on Mars took the vegetable form. Animal races then existing were consciously or unconsciously absorbed into the vegetable races.

The vegetable life, possessed of true intelligence, then evolved into one organic whole in order to obtain the greatest advantage from the limited means of sustenance. Life on Mars is now one vast intellect supported by a vegetable body having its roots in the soil. Such a conception of life resembles the vast being into which the Buddhists say all men will be absorbed.

As the Summer comes on the huge being on Mars stretches its body over the planet, growing in bright orange colored forms. As the heat departs these forms die down and hide their life in the soil till the following season. This produces the appearance of "canals" to us. The reason these canals have such a regular form is that the vegetation follows the lines of regular cracks which occurred in the crust of Mars when it was drying up.

The vast intellect of Mars is occupied with the problems of gaining subsistence from the dying planet and then with investigations of the boundless universe that lies within its sight.

The white spot which we sometimes see on Mars is not a pile of snow, but really an "eye." Supported on a tenuous flexible transparent column, it can raise itself miles above the surface of the planet and watch the operations of its vegetable body at any point.

That the movements of this planetary eye should have escaped observation from the eye is not surprising. The canals on Mars have only been seen by a few astronomers, and many excellent scientists denied their existence for years after they were first observed. They continued to do so until photographs were taken of the canals. The shifting of the "eye" on Mars, already observed by our telescopes, may very possibly have been due to a movement on a transparent neck as described here, the neck itself not being visible.

The "eye" exercises the functions of watching climatic conditions all over its vegetable body, of sending help to parts in need and of conveying external impressions to the great central intelligence. This vegetable body possesses the power of distributing strength to its various parts and of devising new means of extracting nourishment from the soil and atmosphere.

When not engaged in watching the physical condition of its body, the great "eye" makes observations of the earth, sun planets, stars and the whole universe. From its vast side it is able to see more and farther than all the telescopes of our earth put together.

The Martian Was Conceived by H. G. Wells to Be an Octopus-Like Creature Without Bony Structure but Having a Highly Developed Intelligence. Drawing by H. Lanos.

The Pitcher Plant Devouring a Rat, an Instance of Plant Life Possessing Animal Powers.

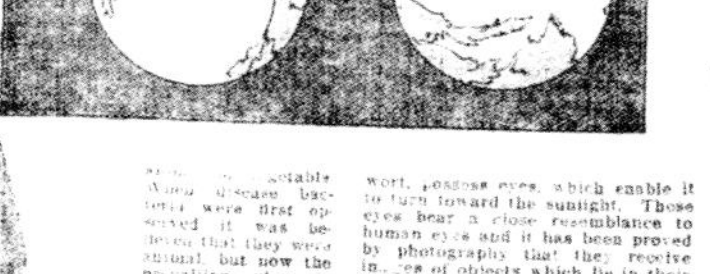

"A vast eye, upon a tenuous, flexible, transparent neck raises itself high above the surface of Mars and can watch the growth of its vegetable body upon any part of the surface."

The Small Diagrams Below Illustrate the Operation of the planetary eye.

The media was right there with Lowell. News about Martians and the possibilities of meeting them was just too exciting to get buried by opposing scientific reasoning. Even the director of the French Botanical Society, Edmond Perrier, chimed in with a few specific thoughts on what our extraterrestrial neighbors might look like. Perrier had no doubt that plants and flowers grew on Mars, and he figured some form of people (surely Martian botanists) must be living there to appreciate them.

In a 1912 interview he explained that due to the "lightness of the atmosphere on Mars and the comparative absence of fierce light, vegetation is luxuriant and the Martians are probably people like the giant Scandinavians." The botanist imagined these giant Scandinavian-like Martians as standing about twice the height of humans, with large heads and bodies, hefty ears, enormous noses, big protruding blue eyes, and white hair. All of these oversized features were supported by rail thin legs and small feet.

How could he be so sure? Much of their physical nature, he assumed, would result from the light force of gravity. It would have allowed their bodies to grow tall, and since it took little effort to walk, their limbs would've remained

lanky. The thin atmosphere would lead to enlarged lungs and chests; the blond hair, Perrier claimed, would be due to the less intense daylight.

Life on the Red Planet was "grand, intense, formidable" for these aliens, since the seasons were much more extreme and the years longer. Animals were similar to those on Earth, but much larger. So were the insects, which were presumably much grosser and harder for Martians to squash with their little feet.

"The year on Mars is twice as long as our earthly one," Perrier explained, "and hence plants and insects have twice the time in which to evolve. Mars is the land of huge plants and ideal flowers, of birds abnormally powerful in song and wondrous in appearance, and of four-footed animals with extraordinarily developed fur and skin."

What's The Matter With Mars?

If all strange theories be true the earth dweller who gets a glimpse of the weird and wonderful Martian populace will stand aghast at a motley collection more grotesque than any monster that ever paraded through dreamland & & &

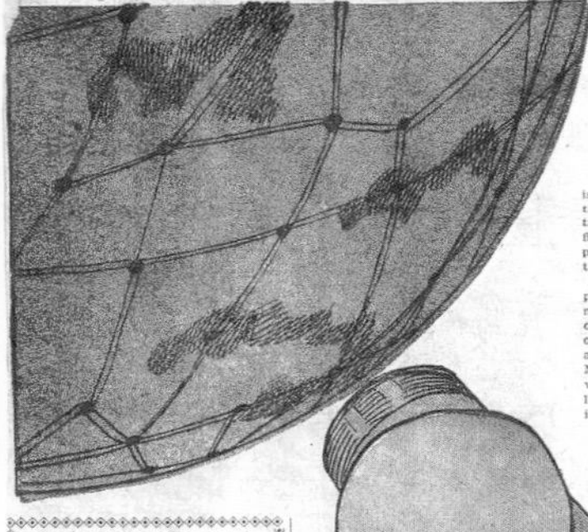

H. G. WELLS' idea of The actual Martian from "The War of the Worlds"
copyright by Harper & Bro.

BY SIR ROBERT BALL.

THAT there may be types of life of some kind or other on Mars is, I should think, very likely.

But what form the progress of evolution may have been it seems totally impossible to conjure. It is true, no doubt, that small planets (like Mars) would be fitted for the residence of large beings, and large planets would be proper for small ones. The Lilliputians might be sought for in a globe like Jupiter, and the Brobdingnagians in a globe like Mars.

MARS and its possible inhabitants are always a source of speculation to scientists, and many and diverse are the opinions as to the little planet and its place in the universe.

For forty years human information as to Mars has gradually grown, but knowledge of the forms of Martian life is mostly theoretic. Percival Lowell, astronomer of the Lowell observatory, believes that life exists on Mars and that the Martian "canals" (so-called) evidence a high degree of intelligence.

"Irrigation unscientifically conducted," he asserts, "would not give us such a truly wonderful manifestation as the several ports as we there beheld. A mind of no mean order would seem to have presided over the system—a mind certainly of considerably more comprehensiveness than that which presides over the various departments of our public works. Party politics, at all events, have had no part in them, for the system is planet wide."

On the other hand, W. S. Holden, astronomer of the Lick Observatory, holds an entirely different view. According to him, the snow caps as seen through the telescope are not composed of snow at all, but rather of solid carbonic acid gas. The lakes and seas and canals are merely color phenomena, such as may be seen upon the moon to-day, and the lines of double canals noted by Schiaparelli, Flammarion, Lockyer and others optical illusions which come from lone straining with the eyes. Between these extremes are all forms of strange theories—people who have an enormous chest development, people who have gills like fishes, people who are formed like lizards or overgrown like giants, four-footed creatures, strong, powerful double-eyed individuals, whose reasoning capacity is always, owing to the age of the planet, thought to be high, and whose social life may be enhanced by material improvements which would make our own look like the implements of barbarism.

In fact, the ideas advanced as to the nature of life upon the fiery planet are so various and so well-sustained that one may readily question whether anything definite is known at all.

One of the most interesting theories concerning the Martians, and upon which all astronomers are apparently agreed, is that if there is any such thing as life in the form in which we know it, it is of enormous size—three to five times as large as anything of the same order here. Men, trees, flowers, birds—all would be of Brobdingnagian proportions, and for the following reasons, which Percival Lowell, the last astronomer of great reputation to reason upon this subject, has given. This is, he says, the effect of mere size of habitat (or the planet on which we live) upon the size of the inhabitant.

"Volume and mass," he says, "determine the force (or downward pull of gravity) upon the surface of a planet, and this affects the size of things. Thus gravity on the surface of Mars is only a little more than one-third what it is on the surface of the earth—a fact which would work in two ways. The first way would act in lightening the weight of things so that three times as much work would be done by expending only once as much or the same muscular force that we do here. In the next place nature could afford to build her inhabitants on three times the scale she does here, since the lightness of the gravitory pull would make them exceedingly nimble. We think of a large person or thing as being exceedingly heavy, but on Mars, where he or I would not have the same weight of gravity to overcome, this heaviness would not exist. Consequently by the mere influence of his size coupled with the greater lightness of the materials with which he would be called upon to deal, he should be able to really do many times as much work as any poor struggling earthling, and to do it with much greater speed."

Now this he scientifically worked out is shown by Professor Lowell, who does not hesitate to draw a very striking picture of the effective powers of the Martian.

"To see this," he says, "let us consider a very simple case—that of standing erect. To this every-day feat is opposed the weight of the body simply—a thing of three dimensions, height, breadth and thickness—while the ability to accomplish it resides in the cross section of the muscles of the knee, a thing of only two dimensions—breadth and thickness.

"Consequently, a person half as large again as another has about twice the supporting capacity of that other, but about three times as much to support. Standing therefore will tire him more quickly.

"If his size were to go on increasing, he would at last reach a stature at which he would no longer be able to stand at all, but would have to lie down. You shall see the same effect in quite inanimate objects. Take two cylinders of paraffine wax, one made into an ordinary candle, the other into a giant facsimile of one, and then stand both upon their bases. To the small one nothing happens. The big one, however, begins to settle, the base actually made viscous by the pressure of the weight above. Now, apply this principle to a possible inhabitant of Mars, and suppose him to be constructed three times as large as a human being in every dimension. If he were on earth he would weigh twenty-seven times as much, but on the surface of Mars, since gravity there is only about one-third of what it is here, he would weigh but nine times as much. The cross section of his muscles would be nine times as great. Therefore the ratio of his supporting power to the weight he must support would be the same as ours. Consequently, he would be able to stand as little fatigue as we.

"Now, consider the work he might be able to do. His muscles, having length, breadth and thickness, would all be twenty-seven times as effective as ours. He would prove twenty-seven times as strong as we, and could accomplish twenty-seven times as much. But to work further work upon mind required, owing to decreased gravity, but one-third the effort to overcome. His effective force, therefore, would be eighty-one times as great as man's, whether in digging canals or in other bodily exercising."

BY HOWARD SWAN.

IN the first place, as regards the physical bodies of the inhabitants of others planets, we cannot tell if their bodies are like our own, or their surroundings.

There may be less or more air there, and so their ears may not register the same sounds. They even may not have any ears; their nerves and muscles under varying conditions of gravitation may be very differently constituted.

But I venture to think that their eyes must be similarly constituted to our eyes, since they live in the same sun's rays, which rays, as we know by experience, can produce the same physical, actinic and electrical effects either with or without air. And further, both in and out of air, fishes, beasts and birds all have eyes.

"As gravity on the surface of Mars is really a little more than one-third that at the surface of the earth—the true ratio is not eighty-one, but about fifty—that is, a Martian would be physically about fifty fold more efficient than men."

Having proved what his physical proportions would be the astronomers are not quite content to rest there, but go on in some instances to set forth a few of his possible physical characteristics, all of which are decidedly interesting. Mr. R. A. Gregory believes "that people with immense chests" or "folk with gills like fishes" could pass a comfortable existence there in spite of the rarefied atmosphere.

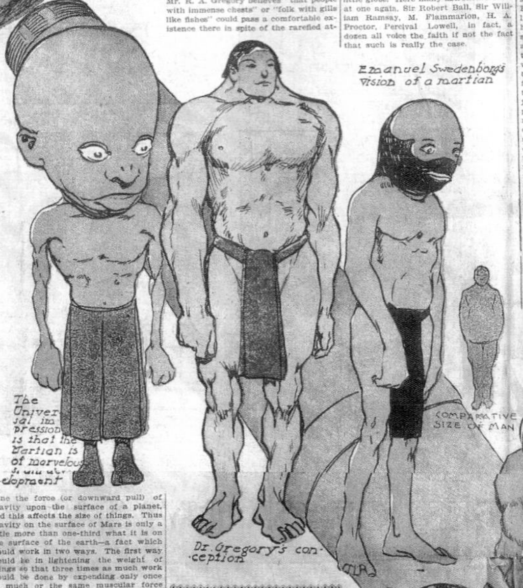

beasts and birds all have eyes." Sir William Ramsay believes that gases or chemical compounds may be intelligent because "it is absurd to suppose that consciousness may not exist with forms of matter the existence of which we are just beginning to suspect."

The most interesting because the most convincing of all this curious argument is that which relates to the possibility of life on Mars—the wonderfully earthlike sphere which swings so far from and so near to our own comfortable little globe. Here many astronomers are at one again. Sir Robert Ball, Sir William Ramsay, M. Flammarion, H. A. Proctor, Percival Lowell, in fact, a dozen all take the faith if not the fact that such is really the case.

Emanuel Swedenborg's Vision of a Martian

Many of them, of course, approach their belief in a very obscure and scientific way, but they approach it, and the pictures which they draw are very alluring. Thus Sir Robert Ball, who stands unquestionably at the head of his profession, draws one of the most pleasing pictures—a picture which has served as much as any other to hearten the modern believer in life upon the little planet and make him feel that some day his faith will be justified.

"That there may be types of life of some kind or other on Mars," says Sir Robert Ball, "is, I should think, very likely. Two of the elements, carbon and hydrogen, which are more intimately associated with the phenomena of life here appear to be among the most widely distributed elements throughout the universe, and their presence on Mars is in the highest degree probable. But what form the progress of evolution may have taken it seems wholly impossible to conjecture. It is true, no doubt, that small planets like Mars would be fitted for the residence of large beings and large planets (like Jupiter) would be proper for small ones. Still I would suggest, however, that as our earth has only been tenanted by intelligent beings for an extremely brief period of its history—say, for example, about one-thousandth part of the entire number of years during which our globe has had an independent existence, we may fairly conjecture that if there is life of the intelligence described, it is not improbable that some method of communication may yet be found, seeing that we are discovering from day to day that which was once the impossible is now the possible and many things that were held are plainly no part of life."

What this means is that life may be there, and it is fair to assume, as Mr. Lowell and many others really do, that it is present now. Mars being old, we know the evolution to its surface must be similarly advanced, and it is highly probably that Martian folks are possessed of inventions of which we have not dreamed. "With them," says Mr. Lowell, "electrophone and kinetoscope are probably things of a bygone past, preserved with veneration in museums as relics of the simple childhood of the race. Certainly, what we see hints at the existence of beings who are in advance of, not behind, us in the journey of life."

And so the investigation of Mars moves forward. Although we have not reached the place where, as some suppose, if we had a flag as large as England or a grove of lights as great as England, we could make them see and understand, still we may be progressing. If there is life of the intelligence described, it is not improbable that some method of communication may yet be found.

HISTORY OF MARS.

B. C. 272—The first known observation of Mars is recorded in Ptolemy's Almagest.

A. D. 1610—The phases of Mars were discovered by Galileo.

1659—The first sketch showing surface details was made by Huygens. He also suggested a rotation of twenty-four hours.

1666—Cassini determined the rotation of Mars to take place in twenty-four hours and forty minutes. He also observed the solar caps, and distinguished on the disk of Mars, near its termination, a white spot advancing into the dark portion.

1777—Sir William Herschel made the first recognizable sketch of the surface detail of Mars.

1783—Sir William Herschel detected the migration in the size of the polar snow caps, measured the polar compression, and determined the inclination of the axis of the planet to its orbit.

1783-1830—Schroter discovered the very dark spots, since shown to be the Northern and Equatorial seas, but supposed then to be clouds.

1840—Beer and Maedler published the first map of the planet, assigning latitudes and longitudes to the various markings. On this map are indicated the first canal and the first of the small lakes.

1862—Secchi made the first study of the colors exhibited by the planet.

1862—Lockyer made the first sketch showing all the forms with which we are now familiar.

1864—Dawes detected eight or ten of the canals.

1867—Huggins detected lines due to the presence of water vapor in the atmosphere of Mars.

1867—Proder determined the period of rotation of Mars within 0.1 second.

1877—Schiaparelli discovered numerous double canals and announced that the appearance formed one of the characteristic phenomena of the planet.

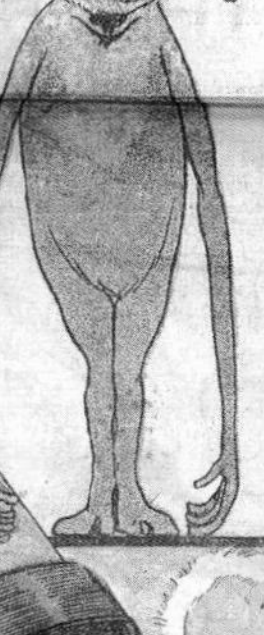

By H. G. WELLS.

A BIG, grayish, rounded bulk, the size, perhaps, of a bear, was rising slowly, and painfully out of the cylinder.

As it bulged up and caught the light, it glistened like wet leather. Two large, dark-colored eyes were regarding me. It was rounded and bad, one might say, a face. There was a mouth under the eyes, the lifeless brim of which quivered and panted and dropped saliva. The body heaved and multifariously pulsated. A lank tentacle appendage gripped the edge of the cylinder. Another swayed in the air.

Those who have never seen a living Martian can scarcely imagine the strange horror of their appearance. The peculiar V-shaped mouth with its pointed upper lip, the absence of brow ridges, the absence of a chin beneath the wedge-like lower lip, the incessant quivering of this mouth, the Gorgon groups of tentacles, the tumultuous breathing of the lungs in a strange atmosphere, the evident heaviness and painfulness of movement, due to the great gravitational energy of the earth—above all, the extraordinary intensity of the immense eyes—culminated in an effect akin to nausea.

There was something fungoid in the oily brown skin, something in the clumsy deliberation of the tedious movements exceedingly terrible.

mosphere," and Howard Swan says that "their eyes are like our eyes." Mr. Lowell suggests that they might be lizards or indeed of any conceivable or unconceivable form, since Mars characteristics "are purely accidental." These views, coupled with those of Sir William Ramsay, who believes that gases or compounds of chemicals, without visible form, might be intelligent; of Swedenborg, the great northern mystic, who saw a Martian "whose face was like the faces of the inhabitants of our earth, but the lower part black, not from a beard, for he had none, but from a blackness in place of a beard," and that of Mr. H. G. Wells, who described his octopus-like visitor as a big grayish rounded bulk," give us a very curious collection of Martians not wholly unsanctioned by science.

Mr. Gregory believes that people with an enormous chest development or gills could live on Mars, because whatever atmosphere exists on Mars must be much thinner than ours and far too rare to sustain the life of a people of our limited lung capacity. Mr. Swan thinks that they have eyes like ours, because "they live in the same sun's rays, which rays, as we know by experience, can produce the same physical, actinic and electrical effects either with or without air, and both in and out of, our air, fishes,

The Universal impression is that the Martian is of marvelous—— ——lant.

The Comparative Size of Man

Dr. Gregory's conception

Another of Gregory's ideas

Sir Robert Ball states that Martians may take this form

One of Dr. Ball's ideas

Professor Lowell asserts that the Martians may be a lizard

WHAT OTHERS HAD TO SAY ABOUT
THOSE FUNNY-LOOKING MARTIANS

This 1904 article offered theories on what Martians might look like from various scientific, philosophic, and creative minds. Most agreed that they were of superior intelligence and therefore had huge brains, as pictured in the leftmost Martian wearing a fez-like hat and high-water pants. Beyond that, ideas shifted dramatically from one thinker to the next, as pictured here.

Professor Richard A. Gregory: The British astronomer believed "people with immense chests" or "folks with gills like fishes" could survive the rarefied atmosphere.

Emanuel Swedenborg: The eighteenth-century philosopher claimed Martians were friendly and humanlike, with yellow faces and black jaws. They lived peacefully without divisive politics.

Sir Robert Ball: The eminent English scientist imagined Martians as having long noses and extremely long, thin arms. These arms, newspapers reported, would allow a Martian to "touch his toes without stooping. Though slender, they would yet be strong enough to enable him to scramble up the front of a house or the top of a tree by merely exerting their pulling power."

Percival Lowell: Lowell was convinced the Martians were giants in stature and mental abilities, but in this article he suggests they might even be lizards or "indeed of any conceivable or inconceivable form." In other words, who knows?

H. G. Wells: As described in *The War of the Worlds*, Wells's Martians were octopus-like with thrashing tentacles. Although his creatures were imagined for a work of fiction, the article notes that his ideas are "not wholly unsanctioned by science."

MEN of MARS AND Other Things

WHAT A VISIT TO MARS WOULD REVEAL — PERHAPS?

A MARTIAN

A MARTIAN ANIMAL ON A SWAMPY GROUND

A MARTIAN FLYING CREATURE

Beyond the physical characteristics of Martians that he described, Perrier also believed they were far smarter than mere humans and had achieved some form of world peace. Like Lowell and Flammarion, he was convinced our neighbors had evolved earlier than our own species and therefore got a big head start in the intelligence department. Surely that meant they had figured out how to do away with war, disease, poverty, and other nonsense, like government and law. We could learn a thing or two from them, and for that reason Perrier wanted to ensure that we Earthlings would be welcoming, not fearful, once communications were established. The moment appeared to finally arrive in February 1920, when Guglielmo Marconi claimed that Martians had reached out to him.

Marconi is better known as the inventor of the radio and also created the first long-distance wireless telegraph. Now he was looking to add extraterrestrial communications to his résumé. The Italian engineer professed to have heard signals on wavelengths seemingly not from this planet. Like Francis Galton before him, Marconi described them as resembling Morse code.

"If there are any human beings on Mars I would not be surprised if they should find a means of communication with this planet," he said. "Linking of the science of astronomy with that of electricity may bring about almost anything. . . . For all we know, the strange sounds that I have received by wireless may be only a forerunner of a tremendous discovery. The messages have been distinct but unintelligible. They have been received simultaneously in London and in New York, with identical intensity, indicating that they must have originated at a great distance."

Not everyone was buying it though. Some simply doubted that Martians could have the ability to send Morse code. How would they learn such a thing? Others thought atmospheric disturbances or sun spots could've caused the bizarre messages.

Tesla, though, welcomed the news and hoped this might be the moment he'd been expecting. "The thing, I think, that we should try to develop is a plan akin to picture transmission, by means of which we could convey to the inhabitants of Mars knowledge of earthly forms," he suggested. "This would enable us to exchange with them not only simple primitive facts but involved conceptions. To talk to Mars seems to me only a matter of electric power and perseverance."

Thomas Edison backed his fellow genius: "Although I am not an expert in wireless telegraphy, I can plainly see that the mysterious wire-

"Hello, Earth! Hello!"

Marconi believes he is receiving signals from the planets

Tesla

Marconi

Niagara Falls

Edison

OF COURSE you recall Jules Verne's "Ten Thousand Leagues Under the Sea." Well, his submarine is now an accomplished fact, isn't it?

And doubtless you read Kipling's "With the Night Mail." Well, the Atlantic has been crossed in a single flight, hasn't it?

Probably, also, you read H. G. Wells' "The War of the Worlds," in which the Martians descended upon us with fighting machines even more formidable than the tanks of the great war and a mysterious agent of wholesale destruction even more deadly than any gas used by either side.

Well, who shall say that Wells hasn't the right idea about Mars being inhabited by beings just as smart as we are—and probably a good deal smarter?

It is a bold man who say "impossible" these days.

Anyway, Guglielmo Marconi, the famous Italian engineer, who perfected wireless telegraphy, has opened up an exceedingly interesting question by this statement:

"I have encountered during my experiments with wireless telegraphy most amazing phenomena. Most striking of all is the receipt by me personally of signals which I believe originated in the space beyond our planet. I believe it is entirely possible that these signals may have been sent by the inhabitants of other planets to the inhabitants of earth.

"If there are any human beings on Mars I would not be surprised if they should find a means of communication with this planet. Linking of the science of astronomy with that of electricity may bring about almost anything.

"While our own planet is a storehouse of wonders, we are not warranted in accepting as a fact the general supposition that the inhabitants of our comparatively insignificant planet are any more highly developed than inhabitants (if there be such) of other planets.

"For all we know, the strange sounds that I have received by wireless may be only a forerunner of a tremendous discovery.

"The messages have been distinct but unintelligible. They have been received simultaneously in London and in New York, with identical intensity, indicating that they must have originated at a great distance.

"These signals are apparently due to electromagnetic waves of great length, which are not merely stray signals. Occasionally such signals can be imagined to correspond with certain letters of the Morse code. They steal in at our stations irregularly at all seasons. We do not get the signals unless we establish a minimum of 65-mile wave lengths. Sometimes we hear these planetary or interplanetary sounds 20 or 30 minutes after sending out a long wave. They do not interrupt traffic, but when they occur they are very persistent.

"The most familiar signal received is curiously musical. It comes in the form of three short raps, which may be interpreted as the Morse letter 'S,' but there are other sounds which may stand for other letters.

"The war prevented an investigation of the Hertzian mystery, but now our organization intends to undertake a thorough probe."

Australia corroborates Marconi's statement. Highly skilled and experienced operators at Sydney have received numerous signals similar to those reported as having been received in England. They consist of frequent repetitions of two dashes, representing the letter M. They are on wave lengths of 80,000 to 120,000 meters. The Australian experts say such wave lengths have never yet been used by any wireless station of the earth.

Now what do the electrical authorities say on the general subject? Here it is, in brief:

Thomas A. Edison has this to say: "Although I am not an expert in wireless telegraphy, I can plainly see that the mysterious wireless interruptions experienced by Mr. Marconi's operators may be good grounds for the theory that inhabitants of other planets are trying to signal to us. Mr. Marconi is quite right in stating that this is entirely within the realm of the possible.

"I have given some thought to the matter and can record one personal experience which may or may not have bearing on proving that Mr. Marconi is right. I was seated on the peak of a great pile of iron ore near the reduction plant at Orange one day, when I noticed that the magnetic needle was jumping about in astonishing fashion. The thought immediately popped into my mind that static signals from some other planet were probably responsible. This idea took such a hold on me that I made the definite suggestion that there be established in the ore fields of Michigan a station where scientific vigil might be kept, in the hope that the great masses of ore in that region would attract magnetic signals from interplanetary space.

"If we are to accept the theory of Mr. Marconi that these signals are being sent out by inhabitants of other planets, we must at once accept with it the theory of their advanced development. Either they are our intellectual equals or our superiors. It would be stupid for us to assume that we have a corner on all the intelligence in the universe."

Nikola Tesla, the famous Serbian inventor and electrical expert says: "Marconi's idea of communicating with the other planets is the greatest and most fascinating problem confronting the human imagination today. To insure success a body of competent scientists should be organized to study all possible plans and put into execution the best. The matter should be directed probably by astronomers with sufficient backing from men with money and imagination. Supposing that there are intelligent human beings on Mars, success is easily within the range of possibility. In March, 1907, I stated in the Harvard Illustrated Magazine that experiments looking to communication with other planets should be undertaken.

"In 1899 I built an electric plant in Colorado and obtained activities of 18,000,000 horsepower. In the course of my experiments I employed a receiver of virtually unlimited sensitiveness. There were no other wireless plants near, and, at that time, no other wireless plants anywhere on this earth of sufficient range to affect mine. One day my ear caught what seemed to be regular signals. I knew that they could not have been produced upon the earth. The possibility that they came from Mars occurred to me, but the pressure of business affairs caused me to drop the experiment.

"The thing, I think, that we should try to develop is a plan akin to picture transmission, by means of which we could convey to the inhabitants of Mars knowledge of earthly forms. This would enable us to exchange with them not only simple primitive facts but involved conceptions. To talk to Mars seems to me only a matter of electric power and perseverance."

Frank Dyson, British astronomer royal, believes we could get Hertzian waves from other planets. Prof. Edward Branley, Paris, inventor of the coherer, is sceptical. Prof. Domenico Argentieri, Rome, says the supposed signals are worthy of careful observation.

Prof. Albert Einstein, the German astronomer and author of the theory of "Relativity" that is apparently upsetting all accepted doctrines, believes that Mars and other planets are inhabited, but if intelligent creatures are trying to communicate with the earth he should expect them to use rays of light, which could much more easily be controlled.

Are there inhabitants on Mars? That's a question on which scientists differ.

Among scientists who have won the right to speak with authority the foremost was the late Professor Lowell, director of the observatory at Flagstaff, Ariz. Not only was Professor Lowell convinced that Mars was inhabited, but he believed the people had a much higher degree of intelligence than those on earth. He dwelt particularly on their inventive genius.

In 1914 he found a new opportunity for strengthening his pet belief by announcing that instead of losing any of their canals the Martians had built two new ones, which could be seen plainly through the telescope.

"We have actually seen them formed under our eyes," Professor Lowell said at the time, "and the importance of it can hardly be overestimated. The phenomenon transcends any natural law, and is only explicable so far as can be seen by the presence out yonder of animate will."

Professor Lowell had little to say about the appearance of the beings on Mars. Edmond Perrier, director of the museum of the Jardin des Plantes, in Paris, constructed the first picture of the Martians as he conceived them. He said in part:

"The men on Mars are tall because the force of gravity is slight. They are blond because the daylight is less intense. They have less powerful limbs. Their large blue eyes, their strong noses, their large ears, constitute a type of beauty which we doubtless would not appreciate except as suggesting superhuman intelligence."

On the other hand, Dr. C. G. Abbott holds that if wireless messages are being received, it is not Mars sending the signals, but most probably Venus. Abbott is director of the Smithsonian astrophysical observatory and assistant secretary of the Smithsonian institution. He says Mars is eliminated as a possibility because known conditions on that planet would not permit the existence of any form of living creature. It is too cold there and there is practically no water vapor in its atmosphere.

Assuming that Mars or some other planet is signaling us, what can we do in the circumstances? Apparently we can do much.

Dr. James Harris Rogers of Hyattsville, Md., who has devoted his life to the study of electric waves and invented the underground and undersea wireless used during the war, declares he is going to undertake to teach the inhabitants of Mars the rudiments of intelligence of this planet. Within a year wireless communication will be established with Mars, Dr. Rogers believes.

L. J. Lesh, a New York radio engineer, suggests that one of the methods of constructing a gigantic station would be to erect huge antennae suspended by balloons like the British dirigible R-34. He asserts, however, that a still better way would be to use huge and brilliant shafts of light as antennae for the system. He thinks that projectors could be grouped around one spot where a great amount of electricity could be generated. He suggests Niagara Falls or some other spot with an enormous amount of water power.

Elmer A. Sperry has a searchlight capable of producing a beam having the illuminating intensity of 1,280,000,000 candle power. He would form a group of 150 to 200 of his searchlights and direct their combined beams in the direction of Mars. An aggregation of that sort would possess the luminous equivalent of a star of the seventh magnitude such as our telescopes are able to pick up readily. Therefore, assuming that the Martians had glasses of equal power, they should have no trouble in catching that dot of light from a distance of 35,000,000 to 40,000,000 miles.

It would be possible, no doubt, to operate these lights so that they could give slow signals which would fill all the requirements of a system of communication. However, an array of lights of this character and the needful energizing plant would cost a pretty sum.

The outlay might be warranted some day, but certainly not until it is certain that we are being called by one of our neighbors out in space.

Georgette and Tricotine.

"My, but you've dyed your hair a fashionable red," commented Georgette.

"Like it?" smirked Tricotine.

"I'll say I do. Is it expensive?"

"Well, yes. I used six bottles of tomato ketchup on the little job."

Fizzless soda water is a fizzle just the same.

An Exception.

"They say Mrs. Bump's table is her weak point."

"Not if you judge it by the butter she puts on it."

Couldn't Be.

"I saw your double on the street today."

"That's impossible, sir. I am a single man."

Philosophical.

Wife—Tomorrow is the anniversary of our wedding day. What shall we do about it?

The Professor (absently)—I suppose we shall have to make the best of it.—London Blighty.

Falling in love is often a serious accident.

Blind Men Splice Rope.

Blind men in a Pacific coast institution have become expert rope splicers and knotters with a very brief series of lessons.

Nail This, Quick!

"Why has he named his motor Wilton?"

"Because it's his car pet."

less interruptions experienced by Mr. Marconi's operators may be good grounds for the theory that inhabitants of other planets are trying to signal to us," he offered. "Mr. Marconi is quite right in stating that this is entirely within the realm of the possible."

Dr. James Harris Rogers, inventor of the underground and undersea wireless system used in World War I, believed that if Martians were sending signals to Earth, they would choose the unusually long wavelength to avoid confusion with terrestrial signals. He felt confident that within a year we'd be talking to them.

Also on Marconi's side was J. H. C. Macbeth, who happened to be the London manager of the Marconi Wireless Telegraph Company, Ltd. Happy to support his boss, Macbeth found the logic to be simple: the wavelengths Marconi received were almost ten times what could be produced at any earthly station, so surely they were from another planet. He, like others, believed the Martians were highly evolved, and so, given their ability to engineer such an intricate canal system, a 150,000-meter wavelength was well within their means.

Macbeth acknowledged that understanding the signals would prove difficult, since no one on Earth spoke Martian. But such minor details were of little concern. "During the war the Germans were able within three weeks to decipher British war messages and we theirs," he said. "No matter what the consonants of the code, no matter what language the message eventually was decided in, whether English, German, Arabic or Siamese, ultimately experts could interpret them."

With technology on Earth rapidly increasing, Macbeth was confident we'd soon be signaling Martians with imagery. "By projection of a lantern slide showing a tree, an operator on Earth, provided we can generate 150,000-meter wavelengths, can follow this with 'tree' repeated many times in international code until the same dots and dashes are repeated by the other planet. Then follow this lantern slide with one showing a man's figure, and repeat 'man' in code. By this Berlitz method you might call it, very shortly it would be possible to surmount language barriers and establish intelligent communication."

FIRST ONE TO TALK TO ALIENS
GETS 100,000 FRANCS (MARTIANS EXCLUDED)

Contacting aliens would be enough of a reward to any scientist. But winning a prize for it would make the whole thing even sweeter.

In her 1891 will, Madame Clara Gouget Guzman bequeathed 100,000 francs to the French Academy of Sciences "to be given, without exception to nationality, to the first person who finds the means of communicating with a star—by this I meant to say a signal to a star and a received response to that signal, exclude the planet Mars because it is sufficiently well-known."

In other words, talking to Martians was deemed too easy. If you wanted the money, you'd have to do a little better than merely establishing communication with creatures more than 35 million miles away.

The Pierre Guzman Prize, as it was known, was created in memory of Madame Guzman's son, who was a big fan of Camille Flammarion and a budding astronomer.

The money was a good incentive for new ideas. One came in 1894 from Frank H. Norton, a writer for the *Illustrated American*. Like Von Littrow and Gauss, he wanted to make a big splashy design on Earth for aliens to marvel at. Of course, his signal would have to reach much farther than Mars. The best place to send it, he believed, was the Great Pyramid of Giza.

"All that there is to do is to obtain the permission of the British Government and to convey to the Pyramid the most powerful electric plants that can be made ready in time," he wrote. "Then run a line of Edison incandescent lights up each angle of the Great Pyramid from base to apex—and you have your signal."

If that wasn't bright enough, he'd added that all the surrounding pyramids could be lit up as well. A nice thought, but not a prize winner.

Hope was renewed in 1920 when Guglielmo Marconi made noise about Martian signals. If Mars was finally in contact, other worlds couldn't be far off, right?

With all his wireless success and Martian studies, Marconi seemed like a good bet to win. His one true rival to the prize was Nikola Tesla. Sure enough, by 1937 Tesla thought he had it in the bag. On the inventor's eighty-first birthday, he told the *New*

Turn Down Scientist's Offer Say He Can Talk To Mars

Paris, Oct. 19.—The French Academy of Science has just turned down the offer made by a German scientist, whose name is not given, who says he is able to get into communication with Mars. At the same time this professor says he is a candidate for the prize of 100,000 francs offered by Madame Guzman to the first scientist of any nationality who succeeds in conversing or communicating with any planet. The German, however, has forgotten that Madame Guzman specially stipulated that the planet Mars was excluded from the competition. French scientists do not believe it is yet possible to communicate with any planet. "Perhaps this will be possible in one hundred years' time," said a member of the Academy. "But the scientist who will succeed in doing so is not yet born, not even in Germany."

This October 22, 1920, article reports on a German scientist who forgot the key rule of the Guzman prize.

York Times that he planned on claiming the award: "I have devoted much of my time during the year past to the perfecting of a new small and compact apparatus by which energy in considerable amounts can now be flashed through interstellar space to any distance without the slightest dispersion."

Tesla planned to submit a full description of his wondrous device to the Academy and told the *Times* he was "feeling perfectly sure that it will be awarded to me. The money, of course, is a trifling consideration, but for the great historical honor of being the first to achieve this miracle I would be almost willing to give my life.

"I am just as sure that prize will be awarded to me as if I already had it in my pocket. They have got to do it. It means it will be possible to convey several thousand units of horsepower to other planets, regardless of the distance. This discovery of mine will be remembered when everything else I have done, is covered with dust."

Tesla did not get the prize. Either he never made the submission, or it never made it to the Academy. Instead, the prize was awarded to the moon landing crew of *Apollo 11* in 1969. Apparently after nearly eighty years of waiting for a winner, the astronauts' giant leap for humankind was good enough for the Academy.

For David Todd, the Marconi news must have been invigorating. Our wannabe hot-air balloonist was ready to do his part to shmooze with Martians. In fact, he'd been spending the previous few years preparing for the 1920 opposition.

This time the professor planned to make the skyward trip from Fort Omaha in Nebraska, on April 23, at which time Mars would be nearer to Earth than on any other day that year. Todd believed that if the Martians were smart enough to communicate with us, this would be their big day to shout hello. So he hired the country's top balloon man to help.

That man, Captain Leo Stevens, was the chief balloon instructor for the U.S. Army, and he happily signed on to build and pilot Todd's inflatable airship. As he had back in 1909, Todd's plan once again was to ascend 50,000 feet or higher. Newspapers reported that they would have "all the facilities of the War Department's chief balloon school, experts from the Rockefeller Institute, apparatus from Johns Hopkins University, the very latest inventions in wireless telegraphy and wireless impulses and specially built instruments of different kinds." How could they go wrong?

High above everything, Todd would use his new gadgets to send out wireless impulses and listen for incoming signals—all of which would be recorded "on a supersensitive plate" of his own invention. As the proposed date approached, the press hyped the big event with great anticipation and expectations of a life-changing day. Sadly, for Todd and his supporters, it wasn't to be. According to a report on April 24—one day after the target date—the flight had to be postponed because "War Department permission for the use of government paraphernalia at the balloon station at Fort Omaha has not been given and until that is forthcoming that experiment cannot be made."

Permission never came.

Newspapers began reporting on Todd's trip months before the opposition.

MARS SHMARS—
IT'S ALL ABOUT VENUS

The world may have been abuzz about Mars, but Nobel Prize winning scientist, astronomer, and professor Svante Arrhenius was sick of the hype.

"The old fellow—dying or already dead—is covered with hard frozen sand and is so dry that its only rain is meteoric dust," he said in 1922. "Perhaps a few sea weeds still exist."

So, why give such a bleak place so much attention? As for the canals, he believed the lines were earthquake fissures.

Not one to pooh-pooh all the planetary excitement, Arrhenius offered an alternative: Venus. The much younger, prettier planet was just waiting to be scrutinized. The constant rain clouds were conducive to life, he believed, and cooler polar regions might be habitable. In a billion years intelligent life might flourish there.

"When the Earth is extinguished," Arrhenius said, "it will be Venus, queen of the heavens, that will take over the role as carrier of culture."

n the meantime, Marconi happily stayed put on Earth and listened for more wireless signals until an explanation presented itself in 1922. To his surprise and surely his disappointment, the source was not Mars. It was the less exotic town of Schenectady, New York—home of General Electric. Marconi discovered the news upon a visit to the company's labs, where one of the researchers admitted to conducting experiments with a wavelength of 150,000 meters. The experiment had been unannounced, and the current had spread across the planet.

This was a mere setback. To keep the Martian dream alive, one simply had to seek a new path, and the relentless David Todd thought he'd found one in 1924. On August 23, with Mars only 34 million miles from Earth, the always-eager professor was once again determined to capitalize on the opportunity. This time, however, he'd do it without a giant balloon. Instead, he used a "radio photo message continuous transmission machine," which had been invented by C. Francis Jenkins. The device would document interplanetary radio signals by producing black lines on a roll of film

A fanciful interpretation of the most promising plan for communicating with Mars. All that is necessary is for Marconi or some one else to find the way to make it possible—"a huge sending station atop towers rising thousands of feet from the surface of the earth; an airplane landing supported in space by balloons; the airplane equipped to relay the wireless waves to Mars." The artist has included a glimpse at Mars as it may be through a magnifying glass.

whenever it received a signal at a 6,000-meter wavelength.

Todd called for hundreds of thousands of radio stations throughout America and Europe to fall silent during five-minute periods starting at midnight and continuing for thirty-six hours.

"Without doubt Mars was inhabited in the past by rational beings, so why not try to stretch hands across the ether? If the Martians ever make attempts to communicate with us, now is the logical time to expect it," Todd reasoned.

The U.S. military agreed to listen for signals but would not comply with the moments of silence. Neither did the commercial radio stations—except for one in Washington, DC.

Still, Todd diligently listened with his machine for twenty-nine hours. It resulted in thirty feet of film, six inches wide, filled with dots and dashes. Always the optimist, Todd interpreted the lines as resembling a face:

> *"The Jenkins machine is perhaps the hypothetical Martians' best chance of making themselves known to Earth. If they have, as well they may, a machine that now is transmitting earthward their 'close-up' of faces, scenes, buildings, landscapes and what not, their sunlight values having been converted into electric values before projection earthward, all these would surely register on the weirdly unique little mechanism."*

Jenkins disagreed. "I don't think the results have anything to do with Mars," he said. "Quite likely the sounds recorded are the result of heterodyning or interference of radio signals. The film shows a repetition at intervals of about a half hour, of what appears to be a man's face. It's a freak which we can't explain."

MARTIANS MIGHT BE MORE ADORABLE
THAN YOU THINK

So, if intelligent Martians weren't signaling us with flashing lights and feasting on vegetation, what kind of creatures *were* living on this far-flung planet? Beavers.

At least, that's what one journalist for *Popular Science Monthly* proposed in 1930. Though most scientists had moved past the idea of canals and extraterrestrial geniuses, there was still belief in vegetation and water on Mars. That being the case, perhaps life originated on the Red Planet the same way it did on Earth.

"The thing to expect on Mars, then, is a fish life much like that on Earth, the emergence of this fish life onto the land, and the evolution of these Martian land-fishes into reptilelike creatures," the article theorized. "Finally, animals resembling the Earth's present rodents like rats, squirrels, and beavers would make their appearance."

The story then suggests that evolution stopped with the beaver because "there seems no reason to believe that Martian life has gone farther than that." Our planet, it suggested, pushed evolution forward with all of its geographical and climate changes. Mars stagnated, leaving the beaver as its ultimate creation. It could live on land or in water, and its fur coat would keep it warm throughout the hundred-degrees-below-zero nights. Despite similar evolution, Martian beavers would surely look different from our version. Bigger eyes would help them better navigate terrain in the weak sunlight, and the lesser gravity and lower oxygen levels might have resulted in larger bodies and chests.

"Herds of beaver-creatures," the article offered, "are at least a more reasonable idea than the familiar fictional one of manlike Martians digging artificial water channels with vast machines or the still more fantastic notion of octopuslike Martians sufficiently intelligent to plan the conquest of the Earth."

Crazy as this "reasonable idea" sounds, it echoed a theory put forth four years earlier by Professor Philip Fox, head of the Dearborn Observatory of Northwestern University. Fox felt sure that vegetation filled Mars, which surely meant animal life would follow. "The Martian animal, however, is probably a fur-bearing one, equipped by nature to live in the wastes around the polar snow caps," he said. "It must be of necessity quite small to migrate rapidly with the changing seasons. It probably would be amphibious, something like our seal, enabling it to swim along with the streams of icy water melted from the snow caps."

Or maybe something like a beaver.

THE MARTIAN TORCH IS PASSED

Percival Lowell left Earth in 1916, but Earl Slipher didn't let the astronomer's unwavering beliefs die with him. A Lowell protégé, Slipher spent decades taking more than one hundred thousand photos of Mars to study its canals and prove their existence to the growing number of naysayers. He published more than five hundred of these photographs in his 1962 book *The Photographic Story of Mars*.

Slipher was one of the last scientists of his era still clinging to early theories. But he wasn't alone in keeping hope alive. A young Dr. Carl Sagan also believed that vegetation and life on Mars was possible. Frozen water in the subsoil, he theorized, could be melted by volcanic activity and create warm pools where life might exist.

As for Lowell's grand vision of Mars, Sagan, along with his collaborator Dr. Joshua Lederberg, professor of genetics at Stanford University School of Medicine, was ready to move on. "The swing of general opinion about Mars has undoubtedly been overcolored by lurid fantasies of canal-building humanoids, which have played no part in serious scientific analysis," Lederberg said. "Now that these have been happily relegated to their proper place in imaginative fiction, our study of the solar system can focus on rigorous factual questions which continue to have the deepest scientific and philosophical interest. Paramount among these is whether life, in any form, has evolved independently of the terrestrial system and man."

Luckily for Sagan and Lederberg's curious minds, they'd soon get a far better look than any of Lowell's predecessors ever imagined. A team of ambitious Soviets would make sure of it.

"**The Martian intelligences might look upon us as we look upon monkeys in a menagerie, and their learned doctors might say: 'See what we were like once! These creatures have a gleam of our intelligence, and their limbs and sense organs indicate the line of evolution that ours have followed.'**"

—**Professor Garrett P. Serviss,**
    discussing what Lowell believed Martians
    may have thought of us in 1916. Lowell's beliefs
    remained steadfast right up to his death.

WHAT DO THE MARTIANS THINK OF US NOW?

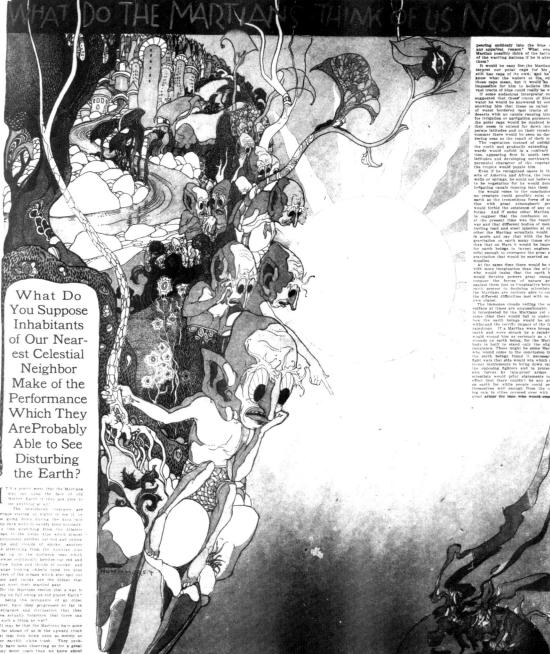

What Do You Suppose Inhabitants of Our Nearest Celestial Neighbor Make of the Performance Which They Are Probably Able to See Disturbing the Earth?

IT'S a pretty mess that the Martians may see upon the face of old Mother Earth if they are able to see anything at all!

The bewildered creatures are perhaps staring up nights to see it, or else going down during the days into deep dark wells to satisfy their curiosity.

A line stretching from the Atlantic Ocean to the Swiss Alps which almost continuously belches out red and yellow flame and clouds of smoke; another line stretching from the Austrian Alps clear up to the northern seas which likewise continually belches out red and yellow flame and clouds of smoke; and strange looking objects upon the blue waters of the oceans which also spit out flame and smoke are the things that must meet their startled gaze.

Do the Martians realize that a war is going on full swing on old planet Earth? Or, being the occupants of an older planet, have they progressed so far in intelligence and civilization that they have actually forgotten that there can be such a thing as war?

It may be that the Martians have gone so far ahead of us in the upward climb that they look down upon us merely as poor earthly white trash. They probably have been observing us for a great many more years than we know about with their powerful and highly perfected telescopes. It may be that they know all that we do and have been doing for many centuries past and knowing this, it may be that they have absolutely refused to give us any recognition until we have reached a degree of civilization equal to their own.

If Mars has inhabitants there is no question but that these inhabitants have reached a higher degree of civilization than earth beings possess, and that, therefore, all their tribal, national and racial wars were fought out long ago. Their interest, surprise and disgust towards earth at this time must be tremendous. If they

have any word worse than "barbarians" they have unquestionably been hurling it down at us night and day for the past five months.

The Martians undoubtedly have better instruments for observing us than we have for observing them. But for every peculiarity we encounter in interpreting the markings on that planet they would have a dozen peculiarities in interpreting the features on the surface of the earth.

Nearly everybody who has investigated

the subject of life on Mars has come to the conclusion that if there is any life there it is also a higher degree of intelligence because it is an older planet than the earth.

Yet the intelligence has been associated for ages with a planet having only slight elevations of land, a very thin atmosphere, a scarcity of water where ice has been used for ages through artificial channels, having vast tracts of deserts and within these deserts large oases fed by forgetting

canals, regions of sparse vegetation and no large bodies of water.

With these conditions going beyond the history of the present inhabitants what must the Martians think of the surface features of this world and the strange things that are now taking place upon it?

If a Martian can observe the earth as we see large red and yellowish areas, extensive greenish areas and besides large regions of varying shades of white, with strange flashes of flame and puffs of smoke bursting from them and with many of these objects disappearing

pearing suddenly into the blue without any apparent reason? What could the Martian possibly think of the battle fleets of the warring nations if he is able to see them?

It would be easy for the Martian to interpret our polar caps for his planet still has caps of its own, and he would know what the waters at the edges of those caps mean, but it would be almost impossible for him to believe that such vast tracts of blue could really be water.

If some audacious interpreter on Mars suggested that these tracts of blue were water he would be answered by some one showing him that these so called bodies of water bordered vast tracts of sandy deserts with no canals running into them for irrigation or navigation purposes. Even the polar caps would be doubted because they seem to extend far down into temperate latitudes and on their recedence in summer there would be seen no dark bordering seas as the result of their melting.

The vegetation instead of unfolding at the north and gradually extending southwards would unfold in a contrary direction, appearing first in south temperate latitudes and developing northward. The perennial character of the vegetation in the tropics would puzzle him.

Even if he recognized oases in the deserts of America and Africa, the results of wells or springs, he could not believe them to be vegetation for he would detect no irrigating canals running into them.

He would come to the conclusion that no creature could possibly exist on the earth as the tremendous force of gravitation with great atmospheric pressure would forbid the existence of any organic forms. And if some other Martian were to suggest that the confusion on earth at the present time was the result of a war and that different bodies of men were hurling lead and steel missiles at one another the Martian scientists would laugh in scorn and say that with the force of gravitation on earth many times stronger than that on Mars it would be impossible for earth beings to invent engines powerful enough to overcome the great pull of gravitation that would be exerted on these missiles.

At the same time there would be others with more imagination than the scientists who would insist that the earth beings would develop powers great enough to conquer the forces of nature arrayed against them just as imaginative beings on earth protest to doubting scientists that the Martians are entirely able to conquer the different difficulties met with on their own planet.

The immense clouds veiling the earth's surface at times are unquestionably rightly interpreted by the Martians yet at the same time they would fall to understand how the earth beings would be able to withstand the terrific impact of the falling raindrops. If a Martian were brought to earth and were struck by a raindrop it would wound him as seriously as a bullet wounds an earth being, for the Martian's body is built to stand only the slightest resistance. There might be some Martians who would come to the conclusion that if the earth beings found it necessary to fight wars that also would with which could invent instruments to bring down rain on the opposing fighters and to protect its own forces by rain-proof armor. The scientists would print statements to the effect that there couldn't be any armies on earth for while people could protect themselves well enough from the crushing rain in cities covered over with rainproof armor the men who would compose

these armies would have to have tons of iron and a crust like a turtle to be impervious to the rain that might crush them anywhere in the fields.

Another feature of earth life which would probably lead the Martian scientists to believe that our world is impossible is the widely different seasons. Declaring that Mars is perfectly balanced as to temperature, he reasons that the earth being so much nearer to the sun would be far hotter in summer for its army to exist in the open and so much colder in winter and so in winter.

The yellow areas he would interpret as desert land. The greenish areas he might consider as vegetation. But what would he make out of the large regions of blue? This world certainly puzzle him because, unfamiliar with oceans, he could not believe that such vast tracts could really be water. And what would he possibly think of the strange objects moving about on these blue areas, singly and in groups, now standing still, now rushing this way and that, with strange flashes of flame and puffs of smoke bursting from them

CHAPTER 2
SATAN, THE SPACE RACE, AND MARTIAN ROBOTS

Russia got a head start in the Space Race with the launch of *Sputnik* in 1957, but America's space program wasn't far behind, thanks to some early help from an unexpected ally: the Devil.

The engineers who developed America's spacecraft didn't sell their souls in exchange for knowledge, but the beginnings of the NASA Jet Propulsion Laboratory (JPL) in Pasadena, California, are, oddly enough, rooted in the occult. It all started back in the 1920s, when co-founder John Whiteside "Jack" Parsons was a teenager and began exploring black magic by summoning Satan in his bedroom. He succeeded only in spooking himself. Parsons later wrote about the experience in the second person, calling it a "magical fiasco" that "was needful to keep you away from magick until you were sufficiently matured."

In a less disturbing venture, Parsons also began exercising his imagination through the adventure stories of Jules Verne and pulp magazines, like *Amazing Stories*. He and a close friend named Edward Forman tried to emulate what they were reading about by heading to the desert and experimenting with crudely made rockets, discovering the finer points of Isaac Newton's Third Law of Motion. That law, if it's been awhile since high school physics, states: "For every action, there is an opposite and equal reaction." In simplest terms, when chemicals are burned in a combustion chamber, the hot gases push out the back end and, in turn, push the rocket forward.

To some, their antics were mischief, but to Parsons and Forman, rocketry was a passion. After high school, the two began to advance their experiments and soon needed more chemicals and more help in the mathematics department. They found both at California Institute of Technology, where they hooked up with an engineering student named Frank Malina. Despite initial hesitation from the powers that be, the aspiring group got permission to work on a thesis project focusing on liquid- and solid-fueled rockets. Up to that point, the whole idea of rocketry was considered nothing more than Buck Rogers nonsense. Parsons, Forman, and Malina didn't care if their ideas sounded like science fiction. Armed with full access to the facilities at Caltech, they would forever change people's perceptions about what was possible. Along the way, as the young rocketeers flirted with explosives, they became known as the "Suicide Squad" (long before DC Comics used the term).

The young men who would launch JPL, on a break. From left to right: Rudolph Schott, Apollo Milton Olin Smith, Frank Malina, Ed Forman, and Jack Parsons. November 15, 1936.

AROUND THE WORLD WITH VERNE'S DISCIPLES

Parsons and Forman were hardly the only ones inspired by the writings of Jules Verne. Konstantin Tsiolkovsky, Robert Goddard, Hermann Oberth, and Wernher von Braun were all enamored with the French author's ideas and made critical advances in rocket technology that helped bring Verne's stories like *From Earth to the Moon* to life.

Tsiolkovsky, a Russian rocket scientist, published *The Exploration of Cosmic Space by Means of Reaction Devices* in 1903. But his ideas were largely ignored because he was simply too ahead of his time. (See below for his sketch of a greenhouse in space, drawn for his 1933 "Album of Space Travel.")

A few years later, American inventor Robert Goddard, began his own quest to turn Verne's ideas into reality. By 1907 he'd made some noise by firing a powder rocket in the basement of a physics building in Worcester, Massachusetts. In 1914 Goddard received

U.S. patents for a liquid-fueled rocket and a two- or three-stage rocket that used solid fuel. He successfully built and tested the first liquid-fueled rocket in 1926. Reaching a height of forty-one feet and traveling a distance of 184 feet, it was a start.

Around that same time, German physicist and engineer Hermann Oberth also began dabbling in rocketry. He designed the prop spacecraft used in Fritz Lang's film *Frau im Mond* (*The Woman in the Moon*) and then built and launched a real version as a publicity stunt. As his studies evolved, he mentored Wernher von Braun, who also read all of Goddard's work and soon became one of rocketry's biggest names. Von Braun worked with the forces of evil in World War II Germany but later helped Americans develop the giant *Saturn V* rocket that made it possible to put men on the moon.

With Verne's stories continuing to inspire imaginations, who will help us take that journey to the center of the Earth?

Parsons's focus on rocket science was matched only by his interest in the occult. He and his wife, Helen, joined a local group called the Ordo Templi Orientis (OTO), which was headed from afar by Aleister Crowley in London. Crowley, who called himself the Great Beast 666, was dubbed "The Wickedest Man in the World" by the English press. He had brought his Thelemic religion to the OTO in the 1920s and quickly climbed its ranks. His motto was "do what thou wilt."

As a rising star in the cult, Parsons did what he wilt, throwing wild

Jack Parsons stands above a Jet-Assisted Take Off canister at JPL's test site in Pasadena, 1943.

sex parties with drugs, loud music, and naked pregnant women dancing outside. As he continued to fall deeper into the pits of his own hell, he also made great progress with Malina, Forman, and other non-occultist friends. The Suicide Squad evaded death and found success with their solid rocket fuel experiments, and the government rewarded them with funding to continue advancing their studies. Officials were so pleased with their breakthroughs that they didn't bother with concerns about Parsons's unusual behavior, like dancing and chanting Crowley's devilish "Hymn to Pan" while firing off test rockets. To them, the young prodigy was simply a "delightful screwball."

As World War II raged, the U.S. military stepped up its order for guided missiles that could drop more than a thousand pounds of explosives. The Suicide Squad had grown into a much larger team to meet the demands, which soon led to the establishment of NASA's Jet Propulsion Lab in 1943.

Parsons's role diminished as the rocketry program expanded. He became even more occupied with the OTO, where he worked to create real magic under the guidance of Crowley. During this bizarre spiritual and physical journey, he met a charismatic young writer who was highly knowledgeable in the world of magic and enjoyed reveling in the cult's debauchery. His name was L. Ron Hubbard. The two became close friends, and Parsons was eager to let Crowley know of his prized recruit.

"I deduced that he is in direct touch with some higher intelligence," he raved in a letter to the Great Beast. "He is the most Thelemic person I have ever met and is in complete accord with our own principles." Crowley disagreed, thinking Hubbard, the future father of Scientology, was nothing more than a charlatan. Regardless of Crowley's prescient disapproval, Parsons and Hubbard remained tight. The two even engaged in rituals designed to impregnate a woman through "sex magick" so she could give birth to a demon child. (It didn't work.)

Like many of the early rockets he launched, Parsons's career eventually came crashing down. The Cold War and Red Scare led to an FBI investigation of his potential ties to Communist groups and his unorthodox lifestyle. When the Feds discovered he'd been planning to hand over rocket plans to Israel in exchange for a new life abroad, his security clearance was revoked. After Hubbard scammed him out of thousands of dollars, Parsons suddenly found himself out of money and out of JPL. He scraped together cash from odd jobs and sales of bootlegged nitroglycerin and continued hosting occult parties and studying rocketry at home. On June 17, 1952, Parsons was rushing an order of explosives for a special effects company when his work went awry and an unexpected explosion took his life. He was thirty-seven.

Parsons never had a chance to see his work emulate the ideas of his boyhood hero, Jules Verne. War was never the goal. As George Pendle put it in his biography, magic and rocketry "promised to satisfy Parsons's desire to escape the earth spiritually and physically." Had the leader of the Suicide Squad not accidentally killed himself, he might've lived to see the establishment of the National Aeronautics and Space Administration in 1958. NASA absorbed JPL and gave birth to the fantastic world of space travel that Parsons had dreamed of since childhood. By 1961, Alan Shepard, Jr. had become the first man to fly in space and John Glenn orbited the Earth a year later.

Like it or not, space travel was made possible largely due to the work of one very clever, curious, and determined Satan worshiper. God bless America.

HELLO, MARS!

All the hopes and hoopla about Martians trying to contact us had gone nowhere. If life existed on Mars, we'd have to go there ourselves and find it. So, on November 28, 1964, we did.

On that day, NASA JPL's *Mariner 4* spacecraft lifted off from Cape Canaveral and did what six other attempts, including four by Russia, had failed to achieve. Traveling at nearly 10,000 miles per hour and powered by four massive solar panels, the unmanned octagonal ship flew 325 million miles through space, chasing Mars's orbit. By July 14, 1965, a year before primetime television aired in color, humankind had placed a camera within nearly 6,000 miles of the Martian surface. Images from the craft took eight hours to zip back to Earth.

Just two years after Earl Slipher's book showcased his life's work, twenty-two photographs taken in twenty-four minutes from NASA's miraculous machine put all the grand ideas about networks of canals, heaps of vegetation, brilliant Martians with superhuman digging skills and other theories to rest. Close-ups showed none of what Slipher, Lowell, and the others thought they'd seen. Instead, with its pock-marked surface, Mars looked an awful lot like the moon. The *New York Times* went so far as to proclaim it "The Dead Planet." Such

lifeless images may have been a letdown to some, but many scientists still found reasons for excitement. What the pictures appeared to show was a planet untouched by time.

"The remarkable state of such an ancient surface leads us to the inference that no atmosphere significantly denser than the present very thin one has characterized the planet since that surface was formed (perhaps two to five billion years ago)," said Robert B. Leighton, Caltech physicist and leader of the *Mariner* image-processing effort. That meant Mars may still hold clues about the origins of the universe—clues that had disappeared from the Earth long ago.

A dual flyby mission followed soon after, undertaken by *Mariner 6* and *Mariner 7*. The two spacecrafts captured more imagery of cra-

NASA's *Mariner 4* was the first satellite to take pictures of another planet from a close distance.

ters and atmospheric data at south polar regions, but not a single sign of life. (If you're wondering about *Mariner 5*, it went to Venus—and confirmed it was hot there and lacking in Venusians. *Mariner 8* was a bust.)

In 1971, *Mariner 9* became the first spacecraft to orbit another planet. It got closer than ever to Mars and saw things its predecessors hadn't, most notably the presence of channels carved by water. They weren't Lowell's canals, but they sure looked like Schiaparelli's *canali*, and that was a lot more stimulating than the celestial carcass that *Mariner 4* had reported. Nikola Tesla surely would've marveled at such technological achievements but been disappointed to find he had no equals out there. The missions didn't rule out life on Mars, but if any did exist, it was far from what scientists had been imagining. In the meantime, Russia sent two unmanned space probes that same year, *Mars 2* and *Mars 3*, but both failed.

Reaching Mars was an extraordinary accomplishment, yet all the Space Race glory of the sixties went to a mission much closer to home: the moon landing. Add Neil Armstrong's lunar footprint to a few Mars flybys, and by the end of the decade the United States had surpassed Russia's impressive scorecard, which included the first male, female, canine, and tortoise cosmonauts in space.[6] But there was no time for gloating. NASA was ready to combine its two feats and get on with the bigger picture: interplanetary flights, starting with landing a spacecraft on Mars. That began with the *Viking* program in the 1970s, which NASA hoped would lead to putting the first astronauts on Mars by the eighties. The former it achieved; the latter is still a work in progress.

---

[6] First dog in space: Laika, *Sputnik 2* (1957). First man in Earth's orbit: Yuri Gagarin, *Vostok 1* (1961). First woman in space: Valentina Tereshkova, *Vostok 6* (1963). First tortoises to the moon: Unnamed, *Zond 5* (1968).

"Had the *Mariner 4* space vehicle been directed at the Earth rather than at Mars and roughly twenty photographs of no better than 1 km resolution acquired, no sign of life, intelligent or otherwise, would have been discerned on Earth."

—Dr. Carl Sagan,
keeping hope alive in 1970

WELCOME TO MARS. POPULATION: 2.

"If there are beings on Mars, they now know what an Earthling looks like. He has three metal legs but can't walk. He has one long retractable arm with a scoop at the end. He has an enormous head covered with curious projections which apparently serve as eyes and ears. And his name is *Viking 1*, but he doesn't answer to it. In fact, he is downright unsociable."

The *Chicago Tribune* gave an apt description of how the *Viking 1* lander might have looked to intelligent Martian life, had it existed as scientists once believed. But for the hundreds of people gathered at NASA JPL in Pasadena, California, and the thousands of contributors at other institutions, there were no worries about how the spacecraft appeared in the early morning of July 20, 1976. Just as long as it appeared.

"Touchdown! We have touchdown!" a flight controller shouted after receiving the signal from the lander.

Stunned silence filled the room before it erupted with cheers, moist eyes, and corks shooting through the air, thrust by champagne propulsion.

The event capped an eleven-month journey spanning 460 million miles. *Viking 1*, which included an orbiter in addition to the lander and launch vehicle, had raced through space at a clip of 1.3 million miles a day, or 54,000 miles per hour. If there was any disappointment, it was that the landing came a few weeks late. The goal was July 4—to serve as a birthday present for the U.S. Bicentennial—but Viking project manager Jim Martin wasn't pleased with the landing options available on Independence Day and took on the added job of telling the president they needed more time. Engineering outweighed politics.

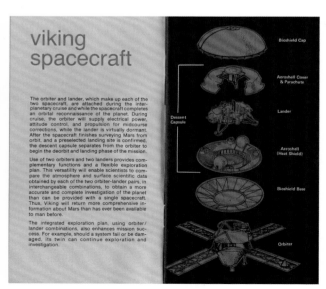

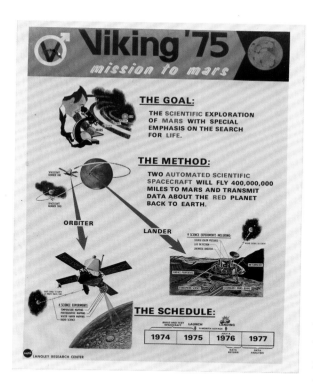

After two more weeks of circling Mars, the *Viking* team found its spot and the lander separated to begin its descent. During this final phase, it measured the distance to Mars's surface to adjust its rate of deceleration and terminate its engines, making it the first autonomous vehicle. Unlike *Apollo*, *Viking* had no astronauts aboard to handle the intricacies of a landing. But by successfully piloting itself, the lander safely reached the surface and began its numerous experiments, including the hunt for life.[7] The orbiter did what its name suggested and continued journeying around the planet to image its surface, measure atmospheric water vapor, and serve as a communication relay for the lander.

Even though the hopes and dreams of so many had been squashed by the early *Mariner* missions, life is resilient, and so were the beliefs that some form of it might still exist on Mars. Microbes could be squirming around in the soil, oblivious to the alien robot parked on top of them. If so, *Viking* was prepared to detect them with three separate experiments. One involved a special broth called "chicken soup" that would act as food for any tiny creatures. If it was ingested by a Martian microbe, the thought was that its metabolism would produce and/or consume gases, like oxygen, hydrogen, methane, and carbon dioxide. Any such atmospheric change in the test chamber would indicate life. The second experiment would create a Martian-like atmosphere in the test chamber, label it with radioactivity, and test whether or not light would be photosynthesized into organic matter by microorganisms. The third experiment would disperse radioactive food into a soil sample and test for exhaled radioactive gas. Scientists would analyze the results and determine if they showed signs of life.

Carl Sagan, however, wasn't convinced that life had to be so puny. Microbes would be great, but macrobes were even better. "There is no reason to exclude from Mars organisms, ranging in size from ants to polar bears," Sagan said in 1975, more than a year before *Viking 1* landed. "And there are even reasons why large organisms might do somewhat better than small organisms on Mars."

That may sound outlandish, but there was a time when sending a robot to Mars did, too. Sagan figured that larger creatures could retain more heat, which would aid survival on a planet that dips into the minus-160s Fahrenheit at night. Water could be found by munching on rocks or eating ice. To survive the daily

---

7 Technically speaking, scientists were searching for metabolic responses based on their understanding of life as we knew it then.

attack of deadly ultraviolet rays, they might develop protective exoskeletons, like insects. If Sagan was right, he wasn't about to miss it, so he worked with engineers at Martin Marietta and other subcontractors to develop camera experiments that would ensure *Viking*'s readiness for larger creatures that may stroll by.

Well, mostly readied. In the original *Viking* designs, a major light was to illuminate the surface in front of the lander at night, but the plan was nixed for financial reasons. Dr. Gentry Lee, who served as *Viking*'s Director of Science Analysis and Mission Planning and is currently NASA's Chief Engineer of the Solar System Exploration Directorate, recalled that his colleague was not happy about the decision. "Sagan filibustered, saying, 'Oh my goodness, you realize what a terrible mistake we're making, what if all the creatures are nocturnal and in the morning we'll wake up and see our pictures and see little tracks in front of the lander?'"

Before *Viking 1* ever got its three metal feet off the ground, scientists had to make sure it wouldn't end up destroying any Martian critters it might find. Anyone who approached the spacecraft before launch had to be clean. Biologically clean, surgically scrubbed. Germs on the *Viking*—even those perfectly harmless on Earth—could hitch a ride and possibly kill off Martian life or confuse the search for it. The team of engineers wore decontaminated clothing and surgical masks to fine-tune every aspect of the craft. Just before liftoff, *Viking* took its final precautions and spent three days being sterilized in scorching 230°F heat. Scientists and engineers had to develop new ways to create electrical components, valves, lubricants, and alloys to withstand the demanding

process, not to mention the bitter temperatures they'd soon face on Mars.

The results of all these intense preparations trickled in nineteen minutes after the lander reached the surface and started its camera. That's how long it took for the first pixels to travel more than 200 million miles back to NASA JPL, where the picture appeared on the screen as a single vertical strip about two inches wide. The image formed one dramatic strip at a time until it was complete, twenty-two minutes later. Finally, after millennia of gazing at the Red Planet with naked eyes, fancy telescopes, and a flyby camera shooting from thousands of miles away, humans were looking at a detailed close-up of the sandy, rocky Martian surface—along with a foot from the lander. Considering the massive budget of the mission, it was truly a billion-dollar photo. That investment immediately offered an important piece of information: the surface was solid.

"People ask why the first photograph was of the footpad, which seemed like a pretty stupid photograph," Lee says. "But I remind people that one-third of the scientists thought the consistency of Mars would be like that of Rapid Shave and go glug, glug, glug and disappear down."

With that mystery cleared up, *Viking 1* continued sending photos as it probed the soil and began its search for life within it. On September 3, 1976, it was joined by another big-headed, three-legged metallic Earthling named *Viking 2*, also equipped with cameras and life detection experiments. The two orbiters and landers continued sending photos until the last of them was retired in 1982. If life existed, it evaded their best efforts—at least according to many NASA scientists and

Work is underway on the *Viking* program, 1974.

astrobiologists elsewhere. However, some believe *Viking* found evidence that says otherwise.

Remember those three biology experiments that *Viking* brought to Mars? One of them tested positive. On sol 8, *Lander 1*'s sampler arm reached out, scooped up a soil sample, and four days later fed a portion the radioactive nutrients. The experiment, designed by Dr. Gil Levin and his co-experimenter Dr. Patricia Ann Straat, was originally called Gulliver. "It was going to far off places to look for tiny beings," Levin explains. "I thought [the name] was cute."

NASA didn't agree and renamed it the much more pragmatic "Labeled Release Experiment," since the potential release of radioactive gas by microorganisms would be labeled with radioactive carbon. Levin and Straat tested the process thousands of times with soil samples from all over Earth and never got a false positive or false negative. Lo and behold, when the Martian soil was analyzed, the results appeared to indicate life. Something had eaten the radioactive lunch. After centuries, if not millennia, of wondering and theorizing about life on Mars, humans had found it.

"When those data points flashed on, I was totally amazed," Straat recalls. "What an incredible surprise that was."

"I was ecstatic," Levin says. "I immediately sent out for champagne."

Everybody else was ecstatic, too. How could they not be? Yet skepticism kept poking at all the exhilaration in the air. No results had come in from the control experiment, which heated another sample to 160°C.

The first image taken on the surface of Mars, with the *Viking* lander's foot seen in the right corner.

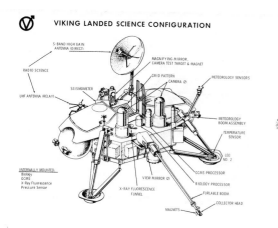

VIKING LANDED SCIENCE CONFIGURATION

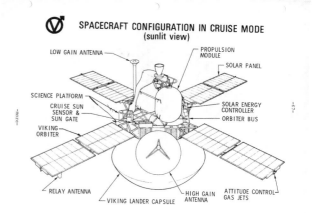

SPACECRAFT CONFIGURATION IN CRUISE MODE
(sunlit view)

The *Viking* lander had many configurations of its science instruments during development.

This temperature was high enough to kill any microbes (at least as they're known on Earth) but not high enough to destroy oxidizing chemicals that might have reacted with the radioactive nutrients and given a false positive. A negative result from the control would ensure that the positive results from the primary experiment came from a biological source. Sure enough, the result came back negative. That seemed to confirm everything—but not everyone was convinced.

"As soon as that happened, there were I don't know how many theories that came out to account for it non-biologically," Straat says. "And so began the controversy, which still rages today."

Months later, the second lander repeated the same experiment on the other side of Mars, four thousand miles away, and tested positive for the presence of metabolism as well. So did repeat tests, all of which gave Levin and Straat even more reason to believe *something* was alive on Mars. Yet others remained doubtful, particularly after a separate *Viking* test, conducted by a gas chromatograph mass spectrometer (its friends call it GCMS for short), looked for organic material and concluded that the soil completely lacked any. This was an unexpected finding considering that scientists knew the Martian surface had been bombarded by meteorites loaded with carbon. It also seemed to indicate that nothing was there that could be metabolizing the nutrients and that the positive results couldn't have been biological. Instead, other members of the team assumed the reaction came from an alien chemical agent. So, the official stance is that the experiments did not detect life. Which meant that when NASA held its first press conference on the testing, Levin could not talk about one of the greatest discoveries in history. In fact, NASA asked him not to.

"All I ever said was our results were consistent with the presence of life," Levin explains. Indeed, he spent many years respecting NASA's request and listening to their theories about why his results could not confirm life on Mars. But after a 1996 announcement about Martian microbes in a meteorite found in Antarctica (page 88), Levin finally began speaking up. His reviews of the test results remained compelling, and other data acquired from Mars and Earth since the *Viking* tests added further support.

Recent analyses of the GCMS instrument have shown that it might have found organic material after all. Its data were re-examined decades later, after NASA's 2008 *Phoenix* lander mission discovered toxic compounds in the soil, called perchlorates. When heated during the GCMS test process, they would've burned up any organic material in the sample. Knowing that, scientists theorized that the reaction would've created chlorinated organic molecules. Sure enough, when revisiting the data, they discovered evidence of chlorobenzene (you know, one of those carbon-based compounds known as a hydrocarbon).

"They're the building blocks of life as we know it—but they're chlorinated," explains Melissa Guzman, PhD student at France's Laboratoire Atmosphères, Milieux, Observations Spatiales (LATMOS), who researched the data with other collaborators. "At the time of *Viking*, this was mysterious to people. They didn't understand why the organics, if they have to do with life, would be chlorinated."

Guzman and her co-researchers have now determined that the chlorinated organic molecules were of Martian origin, but since hydrocarbons had been used as a cleaning solvent for the GCMS instrument, they can't fully rule out contamination from the machine itself. That's why early studies assumed these organics came from Earth.

So, what does all this complex chemistry tell us about biology? Well, the GCMS likely found the organic material that scientists expected to find in 1976, which bodes well for the possibility of life. And while toxic soil may not sound like an ideal home for living creatures, there are organisms on Earth known to metabolize perchlorates. Martian microbes could've simply evolved to withstand it.

None of this has swayed a majority to buy into the Labeled Release results. Adding to the doubt were the negative results from the chicken soup test, called the Gas Exchange (GEX) experiment. But these were problematic because no one could be sure if the gases that came from the soil sample originated from the nutrients put into it. Plus, the GEX had detected "life" in a sample of sterile moon dust. With the Labeled Release experiment, any gas released was labeled with radioactivity, so the scientists knew for certain that it came from the radioactive substrates put on the soil. The only way that could have happened was from some kind of reaction. The third test, known as the Pyrolytic Release experiment, had a small positive result—though much smaller than tests on Earth—and the experimenters were convinced it was a chemical reaction, not a biological one.

NASA has explored various explanations but has yet found an answer that accounts for the positive Labeled Release results. "Nobody has ever disproved Gil Levin's contention that there is no way to explain his results other than some kind of Earth-like life," Gentry Lee claims.

If that's the case, then why aren't we celebrating life on Mars? Levin believes NASA is concerned that if life on Mars is acknowledged, it could delay or impede plans to send people there because we don't know if those microorganisms contain pathogens that might introduce unwanted and unknown diseases into the human body.

"That's the only reason I can think of that makes any sense," he says. "They certainly know, because nothing else has stood up against it. There is no single factor they ever discovered that prohibits life on Mars, or that destroys any aspect of our conclusions."

DR. PATRICIA STRAAT'S NEXT EXPERIMENT FOR FINDING LIFE ON MARS

Straat stands by her results from the *Viking* mission's Labeled Release experiment but acknowledges that life on Mars isn't a fact until we have more information. So, she's come up with another test to help, based on one run by eighteenth-century Dutch scientist Antonie van Leeuwenhoek, aka the "Father of Microbiology." In a nutshell, Van Leeuwenhoek took terrestrial soil and put a little water on it, looked through a microscope, and saw moving things.

"I'd like to do that experiment on Mars," Straat says. But with a twist, as she explains:

> "There's a phenomenon known as cryptobiosis. Nobody knows how long something can live in a desiccated state. It's entirely possible that there are living creatures on Mars that have been in a desiccated state for millions of years. No one knows how long something can remain in that state because we haven't been able to do experiments like that. We know they can remain in a desiccated state for maybe a hundred years. And the phenomenon occurs with creatures more advanced than microorganisms, like tardigrades. They've been considered as models for Martian life. They're surviving radiation as well as desiccation outside the International Space Station.
>
> I'd like to take a little bit of soil, put some water on it, and look at it with a video microscope. First with nothing on it, then with vapor, and then I'd add just a tiny bit of liquid water. I'd watch and see what happens. If you see critters moving, you could also test to see what kills them."

It's similar to a test that was done with the *Viking* experiments. They took soil and put moisture on it. "But we didn't look to see if anything moved," Straat explains. Maybe worsening conditions on Mars over the millennia sent microbes into a cryptobiotic state, as she suggests. If we just add water, maybe we'll find out.

A CLOSER LOOK

When Giovanni Schiaparelli first spotted the *canali* on Mars, he and other astronomers interpreted what they saw as best as they could, and maybe they slanted those interpretations to what they wanted to see. With the photos from *Viking*, things were not much different. Yes, the images were taken from a lot closer, but interpretation remained as wide open as ever, especially if you wanted to believe there was a giant human face on Mars.

Dr. Gerald Soffen, *Viking*'s chief scientist, saw the geological visage—complete with two eyes, a nose, mouth, and some form of head-dress—stretching more than a mile across a rocky mesa in the Cydonia region of the plan-et's northern hemisphere. He reportedly joked with news services about having just seen the first photo of a Martian, but quickly explained that the image is a "trick of light and shadow." He had no intention of launching another wave of Martian madness like the canals had done nearly a century earlier.

A few years later, though, two enthusiastic computer scientists did just that. Vince Di-Pietro and Greg Molenaar were working for a company contracted by NASA's Goddard Space Flight Center in Greenbelt, Maryland, when they uncovered another photo taken from a dif-ferent angle. To them, the face was no joke.

After their work for the day was done, the pair would stick around and use NASA's sophisticated systems to comb through and scrutinize images from the *Viking 1* orbiter. One of those images was of the face (*right*). Thrilled by their find, they started tinker-ing with enhancements and enlargements on NASA's computers and became convinced that a Martian face was staring right back.

Not only that, but other nearby rocky for-mations resembling pyramids raised all sorts of existential questions. Were these remains of a vanished alien civilization? Did Martians visit ancient Egyptians and help build their pyramids? Or did they do a flyby and rip-off Egyptian designs and human faces as soon as they got home? These mysteries sound like nothing more than plots to a science-fiction story, yet the idea of preserving a species by moving to another planet is one reason that space programs are looking to colonize Mars by the 2030s. For now, let's jump back to 1980 when NASA thought DiPietro and Molenaar were just a couple of sensational crackpots.

Conway Snyder, project director for *Viking*, was especially annoyed. "It's utterly ridicu-lous," he said at the time. "Manifestly non-sense. There are a great many more stone faces on Earth than you'll find on Mars, and they weren't carved by men either. It's like seeing images in clouds; it all depends on how the lights hits them."

But plenty of people shared in the amateur sleuths' euphoria, and the publicity that followed only irked officials more. Snyder emphasized that DiPietro and Molenaar had just scratched the surface of what was possible with computer enhancements and that NASA had decades of experience and many more techniques well beyond their comprehension.

"Playing around with computers is one thing; that's great fun—I know how habit-forming it can be—but to go out and represent what you are doing as a new discovery when you don't know the first thing about the field, well, that's just plain stupid," Snyder added.

Snyder felt better after NASA revoked their privileges, ensuring the two computer scientists would never again spend a late night in the equipment room. To their credit, DiPietro and Molenaar never shouted that they had discovered evidence of an ancient civilization. "I'm not drawing any conclusions," DiPietro maintained, "but I'm excited."

If DiPietro and Molenaar were the modern-day Schiaparellis, Richard C. Hoagland was the new Percival Lowell. An amateur scientist, author, and conspiracy theorist, Hoagland took their findings and ran a marathon with them. To him, the mesa was nothing like the mountains or landscapes on Earth. Those, he explained, "are inevitably perspective shots, profiles where if you move a hundred meters one way or the other they go away." The face on Mars, was different because it was "a frontal full-on facial view as if you were looking at yourself in some kind of cosmic mirror."

Hoagland called the surrounding pyramids "the city" and worked out a series of overly complex diagrams with fancy geometry and miraculous measurements to bolster his argument that the structures were purposefully designed. Since this book is not a mathematical guide to everything you ever wanted Mars to be, you've been spared all the nitty-gritty details. But like Lowell, Hoagland preached his findings through interviews and the lecture circuit and put his many meticulous notes into books for all who might be intrigued.

NASA may not have agreed with these amateurs, but it would've liked to. After all, finding evidence of past life on Mars was a major goal of the operation. That said, the Mars Global Surveyor orbiter, launched just over twenty years after *Viking*, explored the entire surface of the Red Planet to help map out ideal landing areas for upcoming rover missions. When it reached Cydonia and took a high-resolution photograph, the face suddenly got camera shy and appeared as if it had covered itself with a veil of natural terrain. The cosmic mirror stared back at hopeful believers, reflecting utter disappointment.

"WHAT *VIKING* NEEDS IS WHEELS."

So said Carl Sagan in 1976. Twenty-one years later, NASA made it happen. And this time, it landed right on schedule to celebrate the United States' 221st birthday.

Remarkable timing, but why did it take two decades after the successful *Viking* mission? Part of the delay was due to a lack of pressure from a scientific community that was resigned to Mars being devoid of life. A lack of money didn't help either. A great portion of NASA's funds went to the Space Shuttle program and, given the enormous amount of information the two *Viking* landers and orbiters provided, the idea of another mission of that caliber seemed unaffordable. The days of billion-dollar projects were over. Yet curiosity remained.

Scientists still had issues to resolve about early Mars and how wet it might have been. Data from *Viking* led them to understand that there were massive amounts of water at one time, but since each lander remained in a single spot, that understanding was limited. If they could follow the water, perhaps it would lead to clues of past—or even present—life. But first, engineers would have to solve a problem right here on Earth: finding a less expensive way to get back to Mars. "Faster, better, cheaper" was management's motto in the early 1990s—a directive intended to cut costs without sacrificing quality.

In most lines of work, you can have only two of those three. But then, most lines of work aren't staffed with teams of rocket scientists. NASA JPL was up to the task. In less than three years and for a mere $150 million, the *Pathfinder* lander and its rover, *Sojourner*, were born.

THE *SOJOURNER* ROVER: BUILT BY ENGINEERING GENIUSES, NAMED BY A TWELVE-YEAR-OLD

NASA is great at building stuff, but sometimes the job of finding the right name requires outsourcing. So, the agency created a contest for students, asking for not only names but also essays to justify their suggestions. Of the thousands of responses that poured in, victory went to twelve-year-old Valerie Ambroise for her submission of *Sojourner*, honoring the nineteenth-century abolitionist and civil rights activist Sojourner Truth.

Children have named *Sojourner's* successors as well, including third-grader Sofi Collis who gave *Spirit* and *Opportunity* their names. Born in Siberia, Sofi was adopted and brought to the United States at age two. In her winning essay, she wrote, "I used to live in an orphanage. It was dark and cold and lonely. At night, I looked up at the sparkly sky and felt better. I dreamed I could fly there. In America, I can make all my dreams come true. Thank you for the 'Spirit' and the 'Opportunity.'"

Sixth-grader Clara Ma named *Curiosity* in 2009. She professed that the word came to her immediately and embodied both her own life and the lives of the thousands of people it took to create the rover. "Curiosity is the passion that drives us through our everyday lives," she wrote. "We have become explorers and scientists with our need to ask questions and to wonder."

DID LIFE ON MARS FALL TO EARTH
13,000 YEARS AGO?

After several missions covering more than a billion miles of space travel, the question of life on Mars appeared to find its answer in a rock from right here at home. In 1984 a team of American scientists on an annual meteorite expedition in Antarctica spotted a keeper in the ice fields of the Allan Hills region. The small potato-sized nugget was the first space rock found during the trip, earning the name ALH 84001.

It wasn't until 1993, however, that scientists determined the meteorite had broken away from Mars some 16 million years ago, after a comet or asteroid smashed into the planet. The meteor took a leisurely jaunt through space for all those years before finding Earth's atmosphere, falling through it, and coming to rest at the South Pole during the Late Stone Age. A team of researchers from NASA and Stanford University discovered that its chemistry matched the Martian characteristics discovered by *Viking*. Then they really dug into it. Over the next two years the group scrutinized every square trillionth of an inch and soon spotted organic compounds produced by biological activity. It all seemed to add up, since the rock was dated to be several billion years old, and therefore had existed while Mars may have been enjoying a warmer, wetter climate—in other words, a period when some type of creature might have crawled into it and fossilized over time.

"It is very difficult to prove life existed 3.6 billion years ago on Earth, let alone on Mars," said Dr. Richard Zare, professor of chemistry at Stanford University, in an August 1996 NASA press release. "The existing standard of proof, which we think we have met, includes having an accurately dated sample that contains native microfossils, mineralogical features characteristic of life, and evidence of complex organic chemistry."

These microfossils had a heavy emphasis on "micro." If what Zare's team found were truly fossils, they were about 1/1,000th the diameter of a human hair. To borrow NASA's analogy: "It would take about a thousand laid end-to-end to span the dot at the end of this sentence."

Tiny as they appeared, they were big enough news to reach the president of the United States. On the morning of NASA's announcement, Bill Clinton stood on the South Lawn of the White House and shared the momentous breakthrough with the world. "Today, rock 84001 speaks to us across all those billions of years and millions of miles," Clinton said. "It speaks of the possibility of life. If this discovery is confirmed, it will surely be one of the most stunning insights into our universe that science has ever uncovered."

He also reminded the public that despite the enthusiasm, the meteorite required further review to confirm the findings. Scientists were doing just that, including many who were irritated that the announcement was made before they had uncovered conclusive evidence. The organic compounds, they believed, could have been formed without the aid of microscopic Martians. Even Carl Sagan, who longed to find extraterrestrial life, admitted that the reported findings were "evocative and very exciting" but were "not evidence of life."

Fossil or not, the announcement rejuvenated interest in the Mars exploration program in advance of upcoming rover missions. The search for evidence of life was just getting started.

After journeying more than 300 million miles, *Pathfinder* burst into the Martian atmosphere at 16,000 mph on the afternoon of July 4, 1997. Its parachute deployed, slowing its fall until it smacked into the dirt at 50 mph. Cocooned within a series of massive bulbous airbags, the lander was perfectly cushioned as it bounced along the surface for about half a mile before coming to a stop right in the region that NASA was aiming for: Ares Vallis (Mars Valley). They picked this spot based on *Viking* orbiter photos that indicated it was an ancient flood plain. It would be a goldmine for Martian rocks, packed with clues about the planet's last few billion years of history. Pathfinder unfolded three solar panels, like a mechanical flower blooming in the insanely cold Martian springtime, and got to work. Six hours later, its first otherworldly photos found their way back to Earth.

The mission's success continued the next day when NASA unleashed its rover. Standing just one-foot tall and weighing a mere twenty-three pounds, the six-wheeled robot could have been a droid straight out of *Star Wars*. It rolled away from *Pathfinder* at a slow and steady pace of half an inch per second, snapping photographs and measuring the chemical and atmospheric conditions of this dusty new-but-very-old world. The first rock it probed, a ten-inch lumpy specimen named Barnacle Bill, was loaded with quartz. This meant it had been heated and reheated multiple times, suggesting that Mars had been warm longer than previously thought.

Designed to last seven sols, *Sojourner* scoured the planet for an impressive eighty-three. By its final transmission, *Pathfinder* and the rover had sent more than 17,000 photos back to Earth, along with extensive data on soil,

Pathfinder and *Sojourner* landed on Mars on July 4, 1997.

> "I FEEL LIKE WE WON THE SUPER BOWL, THE WORLD SERIES, AND THE WORLD CUP, ALL IN THREE DAYS."

—Brian Muirhead,
flight system manager at NASA JPL, in 1997, after *Pathfinder*'s successful landing and *Sojourner*'s foray onto the Ares Vallis surface.

rocks, winds, and weather factors. Ultimately, scientists believed the evidence showed that Mars was once warm and wet, but that was hardly the main goal. NASA proved it could successfully land a spacecraft on Mars, with a rover, on a budget. There were still no conclusive signs of anything living there, but *Pathfinder* and *Sojourner* had given the U.S. space program plenty of life for future missions. This time they wouldn't take twenty years to launch.

At the start of the twenty-first century, NASA began a strategic multi-decade program designed to find evidence of ancient life on Mars. Kicking off the mission were two new rovers, *Spirit* and *Opportunity*, built to study rocks and soils in a search for clues to the presence of water on Mars long ago. With water, life has a chance.

The twin robots nearly doubled the Martian population when they landed weeks apart from each other in January 2004. *Spirit* settled in the Gusev Crater—a 100-mile-wide puncture near the Martian equator that appeared to have a giant water-carved channel emptying into it.

Opportunity landed on the other side of the planet in one of the few small impact craters that dot a relatively flat area called Meridiani Planum. "We've just scored a three-hundred-million-mile-interplanetary hole in one," said Dr. Steve Squyres, principal investigator for the Mars Exploration Rover Project. NASA named the hole Eagle Crater, after the *Apollo 11* lunar module—as in, "The Eagle has landed." Plus, it had a bonus meaning: an eagle is a hole-in-one on a par-three hole in golf.

Squyres, who is also a professor of astronomy at Cornell University, spent years developing proposals for Mars exploration rovers before finally hitting a hole-in-one of his own, scoring both *Spirit* and *Opportunity*. More than twice the size of *Sojourner*, his new and improved robots were designed with sophisticated panoramic cameras and tools to pursue their geological mission.

Opportunity got off to a fast start with photos of bedrock inside its crater. Layers and layers of bedrock. "We're seeing magic," Squyres said of the images as they appeared on the screens at NASA. With the oldest layer being on the bottom and the youngest on top, they offered a record of what happened over different periods of time. "It's like reading a history book, in order," Squyres says. One of its early pages offered a clue with the discovery of jarosite—a mineral known to form only in water.

HELLO LITTLE ROCK,
WHAT'S YOUR NAME?

Scientists observing Mars have named the various regions across the Red Planet, dating back to the first detailed maps drawn by Giovanni Schiaparelli in 1877. Schiaparelli drew from Greek mythology and history to develop his nomenclature, which included such places as Cydonia and Elysium.

More recently, craters have been named after Schiaparelli himself, as well as other curious astronomers who've been instrumental in promoting an interest in Mars. There's Copernicus, Cassini, Herschel, Flammarion, Wells, and Bradbury, to name a few. Then there are other food-related sites, like Cranberry Sauce and Drumstick, which are rocks found by the *Spirit* rover at Husband Hill. They were named after the Thanksgiving weekend in 2004 when they were photographed.

NASA has general rules about naming, such as never honoring people while they're still living. But as the rover team was taking note of interesting rocks during its first few months, it quickly realized that labels like "Rock 1" and "Rock 2" were not very useful. Over the long holiday break, team members programmed the rovers to take a 360-degree panoramic photo. They called the images the Thanksgiving Pan and started naming features after foods associated with the holiday. At first it was all in fun, but they soon realized they'd stumbled upon a useful naming convention.

"If you have all the names of a rock in a given location be united by a common theme it provides this powerful mnemonic that helps you remember what was going on," explains Steve Squyres. "So I see a rock called Cherry Bomb, I think, Cherry Bomb, fireworks, Fourth of July, oh yeah, I remember."

Another *Spirit* site earned the name Home Plate because that's what it looked like from orbit. All the rocks in that area were then named after baseball players. As for the craters, inspiration came from a course that Squyres teaches at Cornell University on the history of exploration. *Endurance* was named for Sir Ernest Shackleton's Antarctica expedition ship in 1912; *Endeavour*, after the 1768 British Royal Navy research vessel; and *Victoria*, honoring Ferdinand Magellan's flagship that circumnavigated the world in 1519.

s *Opportunity* crept out of its impact crater and continued its journey through Meridiani Planum, it discovered something scientists hadn't seen before: little round beads. NASA called them blueberries. They determined that these spheres were made from hematite. Like jarosite, on Earth this mineral forms only in liquid water.

Spirit got off to a slower start, but soon enough it found a little excitement of its own in an area called Columbia Hills, where it uncovered salt deposits and rocks weathered

Tracks left on Mars by *Opportunity*, 2010.

"NO HEADSET, NO JOYSTICK, NO THROTTLE. NOTHING LIKE THAT. IT'S BASICALLY A BUNCH OF GEEKS AT COMPUTER SCREENS SITTING AROUND AND WRITING HUNDREDS OF LINES OF CODES EVERY DAY."

—**Steve Squyres,**
principal investigator for the Mars Exploration
Rover Project, in 2019, on how rovers are driven

by water. It later found evidence of hydro-thermal systems that would have hosted hot springs. Though both rovers were built to last ninety sols, they far exceeded expectations. *Spirit* remained in contact with NASA until May 2011. After getting caught in a Mars-wide dust storm and losing communication in June 2018, *Opportunity* was pronounced dead on February 13, 2019. After nearly fifteen years of exploration, its final resting place is fittingly called Perseverance Valley. Throughout their travels, both rovers provided nearly half a million photos of the Martian surface and hordes of geological data that confirmed the existence of water on Mars billions of years ago, All this new information indicated that at least part of the planet had most likely been a dramatically different place—with water flowing on the surface and warmer temperatures, especially around areas with violent volcanic eruptions. Just the kinds of places that might have been habitable for some form of creatures.

An early 2003 artist concept of a NASA Mars Exploration Rover on Mars. *Spirit* and *Opportunity* were launched later that year.

TWIN ROVERS PAY TRIBUTE
TO THE TWIN TOWERS

On the morning of September 11, 2001, employees at Honeybee Robotics in Lower Manhattan were beginning their day's work on the rock abrasion tool (RAT) for the *Spirit* and *Opportunity* rovers. Dr. Stephen Gorevan, the company's founder and chairman, was on his bike when the first plane struck Tower 1 of the World Trade Center.

In the aftermath of the terror and devastation that assailed the city and the nation, the robotics team wanted to pay tribute to the victims in some way. Tight deadlines prevented them from helping clean up debris, as many New Yorkers did. But, being creative problem solvers with engineering degrees, they figured out another way to honor those who perished. They would send pieces of the tower to Mars.

Working with the mayor's office, they acquired scrap metal from both towers and incorporated it into the cable shield design of the RAT, embedding an image of the American flag in the metal.

"It's gratifying knowing that a piece of the World Trade Center is up there on Mars," Gorevan said. "That shield on Mars, to me, contrasts the destructive nature of the attackers with the ingenuity and hopeful attitude of Americans."

The rovers' findings proved there was much more to learn. So, to continue its studies, NASA built a bigger and better robot. By bigger, we're talking a one-ton mechanical beast the size of a MINI Cooper. As for better, *Curiosity* (*below*) contains more scientific instruments, including a robotic arm with microscopes powerful enough to spot fossils of microorganisms (no luck yet), a laser that zaps rocks, and a chemistry lab that analyzes the vapors from those zapped rocks and tests for the building blocks of life. Though more advanced than its predecessors, it moves slightly slower, cruising at around 0.09 mph.

"It's really a laboratory that happens to have some wheels to move it around, rather than an awesome robotic car that you drive around at 50 mph," says Dr. Ashwin Vasavada, project scientist for *Curiosity*.

The mobile laboratory landed on August 6, 2012, in Gale Crater—a 96-mile-wide hole with a mountain in the middle. This mountain, called Mount Sharp, is more than 16,000 feet high and is built from layered sedimentary rock, each layer made with different minerals. Think of it like Gusev Crater or, closer to home, the Grand Canyon, where each layer was formed over time by wind and water. By exploring these layers and detecting organic materials, minerals, salts, and other elements, *Curiosity* has helped scientists determine that they were formed at a time when the Martian environment may have been habitable.

Even robots take selfies. This self-portrait of *Curiosity* took place on lower Mount Sharp and combines several component images taken by the rover's Mars Hand Lens Imager on August 15, 2015.

JPL WAS HERE

When NASA JPL designed the wheels for *Curiosity*, the engineers wanted to make sure the rover avoided getting stuck in the Martian sands, as *Spirit* had. So, they added three rows of holes to each of the wheels to help clear out accumulating dirt. These holes also created a visual odometer. For example, if JPL programmed the rover to go a certain distance, team members could count the spots in the dirt to gauge how far it went. Practical as the design was, it had yet another purpose that had nothing to do with the rover's performance. It let the world know exactly who put that robot on Mars.

The original prototype had "JPL" written on the wheel among the treads, ready to leave its mark along the surface. But NASA headquarters nixed it, because other NASA centers had contributed to the mission. Hearing "no" just gave engineers another problem to solve. Instead of using letters, they designed the three lines of holes using a series of dots and dashes that spell out "J-P-L" (*below*). Francis Galton and Guglielmo Marconi had once suspected that Morse code was being used on Mars. Well, now it is.

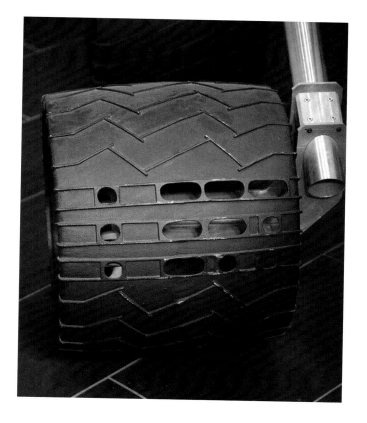

Inside the white box is the *Curiosity* rover and its parachute descending to the surface, as spotted by NASA's Mars Reconnaissance Orbiter.

"We keep getting more and more evidence that it's possible there was microbial life on Mars three and a half billion years ago," says Humphrey "Hoppy" Price, chief engineer for NASA's Mars Exploration Program. "Nothing conclusive, but nothing we've seen indicates that wasn't the case."

The amount of layered rock suggests that if microbial life existed all those billions of years ago, it would've enjoyed habitability for about a million years. That's a blip, but not an insignificant one.

"If you dropped some living organisms on Mars, they would've found a friendly environment to survive in and grow in," Vasavada notes. "The thing that we haven't been able to show is that life made use of that opportunity."

So far, the slow and steady rover has taken its search to an elevation of just over one thousand feet. To help it work a little faster, *Curiosity*'s developers have automated its ability to find samples of interest so it can shoot lasers at will. (If astronauts eventually land on Mars in the same area, they might want to steer clear, for *Curiosity* could live up to its name and zap them to find out what they are.)

Aside from gathering evidence for habitability, *Curiosity*'s roving and zapping have helped it detect cyclical and seasonal spikes of methane. On Earth, methane is largely created by microbes and cow farts. Since the gas fades away, we know it's not lingering from a billion years ago. The puffs *Curiosity* detected are recent. Gil Levin sees it as even more evidence of the life that *Viking* found. But since methane can also come from a hydrothermal process underground, its specific source remains unknown.

Though Levin remains alone in his confidence, bringing back rocks from the Red Planet could help scientists reach a more definitive conclusion. That's the goal for the upcoming sample-return mission, beginning with the next rover: *Mars 2020*.

The Mars 2020 project in development at JPL, 2019.

THE SPHINX, KERMIT THE FROG, AND OTHER AWESOME THINGS FOUND ON MARS

That's right, the Sphinx was spotted on Mars by *Curiosity*. The statue appears to be two hundred feet long and offers proof to conspiracy theorists who've proposed that aliens visited ancient Egyptians thousands of years ago. More amazingly, after Martians helped build the Sphinx, they returned to Earth several millennia later to work with Jim Henson on his design for Kermit the Frog.

Or maybe the people seeing these things are just suffering from pareidolia—the same psychological phenomenon that causes people to see shapes in clouds, as Conway Snyder suggested about the face on Mars. The brain wants to find patterns to explain what it's seeing, and since Mars is so unknown, all those brains staring at its surface have a lot of work to do. Here are a few other sights people have claimed to see:

Beaker: Kermit keeps company with Dr. Bunsen Honeydew's assistant. An image of the Muppet scientist's round eyes and nose, disheveled hair, and perpetually worried mouth was spotted in August 2018 by the Mars Reconnaissance Orbiter's High Resolution Imaging Science Experiment camera (HiRISE) during a dust storm.

A Martian skull: The *Curiosity* rover photographed what appeared to be a bulbous humanoid skull in 2014. ArtAlienTV's YouTube video dissects the finding and calls out the head's nasal cavity, teeth, jawbone, gill-like structures and eyeballs. Yes, eyeballs—not eye sockets. "The eyes may not be like human eyes at all," the video's host explains. "They may have a bony sort of eyes that can survive when the rest of the flesh on the skull has rotted away." An enhanced image highlights the facial details, giving it an uncanny resemblance to the Martians depicted in *Mars Attacks!* Or it may just be a weird rock.

A happy face (left): When the *Viking* orbiter mission photographed Galle Crater, the crater smiled.[8] Two eyes and a mouth curved happily upward. Unlike the face on Mars, no one has claimed this happy countenance proves that intelligent Martians are welcoming us from within the bottom of the hole.

Bigfoot: A humanoid figure photographed by *Spirit* in 2007 appears to have the same shape and posture as the classic grainy Bigfoot footage taken by Roger Patterson and Bob Gimlin in 1967. The evasive creature must have thought he'd finally found a place where he no longer had to hide.

Morse code (below): A series of dots and dashes in an impact crater photographed by the Mars Reconnaissance Orbiter in 2016 excited some observers who believed the design could only be a message created by intelligent beings. Alas, the sand dune patterns were formed by alternating wind flows, not Martians well versed in Morse code.

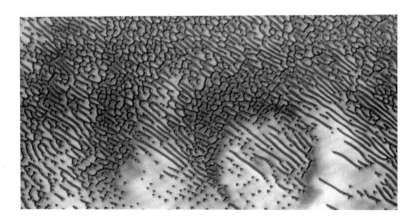

---

8 It's easy to confuse Galle Crater with Gale Crater since one seems like a typo. But it's not. Galle Crater is named after German astronomer Johann Gottfried Galle. Gale Crater is named for Australian astronomer Walter Frederick Gale. Glad we cleared that up.

Much like *Curiosity, Mars 2020* and will be the first step in sending rocks back to NASA. It is projected to land in Jezero Crater (*below*), a pit with a delta leading into it, which indicates it was once a lake. That lake was the size of Lake Tahoe, giving the rover a massive playground to explore.

"It's exciting because, first of all, the fact that there was standing surface water in this lake is attractive to meet our goals with respect to seeking signs of ancient life," explains Dr. Ken Farley, project scientist for the *Mars 2020* mission. "Even better, the delta is—at least among some scientists—thought to be the most likely place that evidence of life could be preserved in rocks. And this is especially true in the very, very fine grain stuff at the bottom of the delta, basically the mud, that was deposited on the bottom of the lake."

On Earth, that kind of mud in similar environments is often filled with organic matter from things that lived in the lake. If we're going to bring samples back from Mars, these are the ones to get.

As the rover begins its collection, it will store the samples in tubes and leave some of them on the surface to be retrieved by a future rover. The rest will be kept in its belly pan. This is NASA's way of splitting the risk: if *Mars 2020* were to keep all the samples, they could be lost if the rover fails or gets stuck; if it leaves them all behind, they may never be picked up if something goes awry with a future mission.

By 2026, in partnership with the European Space Agency (ESA), phase two of the sample-return plan will launch a fetch rover designed to collect the tubes and return them to a Mars ascent vehicle. If all goes well, the *Mars 2020* rover will do the same with its haul. Once the booty is aboard, the ascent vehicle will deliver it to an orbiter, which will space-mail the entire stash back home in an Earth-entry vehicle. Unlike meteorites from Mars that got here all by themselves, scientists will be able to study these samples within the context of knowing precisely where they came from, which will give them a much better understanding of what they're looking at. With a little luck and a lot of science, researchers might finally find a Martian by 2030. Of course, precautions will be undertaken to protect against the possibility of the samples contaminating Earth.

The *Mars 2020* rover will also keep busy with a whole new set of scientific instruments, including an experiment that converts carbon dioxide from the atmosphere into oxygen. "It's very, very small—it's not going to produce enough oxygen to help any human sitting there, so don't expect to go to Mars and take a straw," explains Ann Devereaux, deputy division manager for Systems Engineering and Integration and Test at NASA JPL. "However, it's basically a precursor for something which would be a much larger unit that, say, when Matt Damon goes, they would be able to generate oxygen." That oxygen could also be used to make propellant—so the Matt Damons of the world can make their return trips home.

While the rover makes oxygen and waits for its sample-return partner, it'll be kept company by the first helicopter to fly somewhere besides Earth. This aerial robot will primarily be a technology demonstration designed to show that controlled flight in the Martian atmosphere is possible. In the future, such helicopters could be used to explore places out of the reach of rovers.

Elsewhere on Mars, the rover will be joined by another robot jointly launched by the ESA and Russia. This first-ever European rover, named *Rosalind Franklin* (after the British chemist who helped unlock the mysteries of DNA), is aiming to land in the plains of Oxia Planum, just north of the Martian equator. There, it will begin searching for evidence of past or present life by drilling about six feet into the surface. That's deeper than the most damaging and extreme radiation can penetrate, so its findings promise to offer new answers to Mars's question of life.

For now, the quest to find life on Mars has gotten some help right here at home. Recent discoveries of microbes deep within the Earth, where no life was thought possible, offer evidence that Mars's unfriendly conditions may indeed be conducive to microorganisms. The tenacious little bugs were found by a team of scientists in 2011, after they ventured two miles into the gold mines of South Africa, which happen to be the deepest artificially made holes on Earth. The sun doesn't shine there, but it's plenty hot—and wet, too.

The scientists found that samples of water from rocks formed billions of years ago—and never touched by human hands—were home to tiny worms, crustaceans, and arthropods living quite happily and peacefully in their dark underworld. Life finds a way, even when we think it can't. So, perhaps it can prosper beneath the surface of Mars, where all that ionizing radiation is no longer an issue. The Martian underground might be free of toxic perchlorates and peroxides, and the farther you go, the warmer—and wetter—it could be. Just like in a mine.

"That [environment] is very pleasant if you're a microbe," says planetary scientist Dr. Pascal Lee, who serves as director of the Mars Institute. "So I wouldn't be surprised if there's an entire underground microbial biosphere on Mars."

Lee, who is also director of the Haughton-Mars Project (HMP), has spent decades advocating human exploration of Mars and its two moons. Considering that those years have been filled with extensive research on Antarctica and the Arctic region's geological features as analogues for Mars, you could say he's pretty dedicated to it. A major part of his studies has been Devon Island in northern Canada, the world's largest uninhabited island and home to the Haughton impact crater and ancient river-

The *Spirit* rover created this panorama of the Martian landscape. The site is informally named "Larry's Lookout" and is approximately halfway up "Husband Hill."

beds. Both are similar to those discovered on Mars. These riverbeds, Lee notes, were not carved by flowing water under open air but, rather, by water underneath ice covers sitting on the landscape. If it's possible here on Earth, then it's possible that riverbeds were formed similarly on Mars. That would mean that Mars didn't require a warmer, wetter atmosphere in order to appear as it does today. Meteor strikes and volcanic activity could have sufficiently warmed the ground for liquid water to leave its signature.

"The bottom line is, there's absolutely no requirement from what we've seen of the Martian landscape for it to have been a thicker, warmer atmosphere early in its history," Lee says. "This notion of a planet with oceans that was more earthlike—it's all wishful thinking."

Lee's theory goes against most published papers and the general scientific consensus, but Squyres and others do agree that a warm ground could make for a happy microbial home. Just like the ones hanging out miles beneath Earth's surface. "Those same microbes would probably be perfectly happy on someplace that exists on Mars today," Squyres says. "So the idea that there was once life on Mars and there is, in fact, still life on Mars buried deep below the surface is plausible, but as yet completely untested conjecture."

WELCOME TO
MARS ON EARTH

Before we send humans to Mars, it will help to know what it might be like to explore its freezing cold rocky surface in a cumbersome spacesuit. Fortunately, we can attempt to find out by spending time at the one place on Earth that's similar to Mars: Devon Island. Positioned in the Canadian High Arctic, it's the largest uninhabited island on our planet. It's also home to the Haughton crater—a massive dent spanning more than a mile in diameter, and one of the highest-latitude craters on Earth.

Pascal Lee, director of the Haughton-Mars Project, has spent the last twenty-two years studying Devon Island and running partial simulations with other scientists. Lee calls it "Mars on Earth" because of its geological and climatological similarities.

Part of his research includes developing a spacesuit specifically for the Martian surface. On Earth, the suits weigh 300 pounds, so on Mars it would feel more like 125 pounds. That's just too heavy to schlep around in.

The challenge is, how do you shave off weight without losing the stuff you need to do your job and stay alive? Lee and a team of engineers have been conducting simulations to help them understand the needs of field geology in a Mars-like environment—for example, the frequency of bending down to pick up rocks or the mobility required to reach hard-to-get samples. Then, once you've got those samples, how easy is it to annotate, photograph, and label them? And store them safely in a bag? The team's studies of the various complexities involved in these seemingly simple tasks have helped them design an expanded exploration unit. They've introduced a robotic ATV to the spacesuit system, allowing astronauts to offload weight onto the vehicle. Rather than carrying all your batteries and a large oxygen tank, you'll take smaller batteries and a smaller tank sufficient for a short walk to gather or document rocks. Wherever you go, the ATV goes. "It follows you like a good dog," Lee says.

Future explorers on Mars will want all the help they can get.

"While the moon has been the focus of our efforts, the true goal is far more than being first to land men on the moon, as though it were a celestial Mount Everest to be climbed. The real goal is to develop and demonstrate the capability for interplanetary travel."

—Dr. Thomas O. Paine,
head of NASA, May 26, 1969

More than a century after Nikola Tesla suggested subterranean life, his theory may soon be proven right. All we have to do is look. Lee wants to start by peering inside the dark crevices of a volcano. They may still be active even though they're not erupting and, if so, a warm, wet environment could be flourishing inside—free from radiation. Having already put numerous robots on Mars, taking a peek sounds relatively simple, but rovers aren't ready for that sort of adventure just yet. In fact, rovers may never be ready for it, since taking them to places that might be habitable is restricted by the same Outer Space Treaty that led to *Viking*'s rigorous decontamination efforts.

"Over the years the use of electronics and plastics and all that sort of stuff has made it harder and harder to sterilize simply by heating," says Farley, explaining why replicating the *Viking* treatment doesn't quite suffice. "If you bake our spacecraft, you don't want to bake it too warm, otherwise you have a pile of melted plastic."

So, to go volcano diving, the best way, according to Lee, is with a drone hovering over the interior. Not the kind you got for your birthday, the kind that is currently being developed with rocket thrusters to combat a thin atmosphere. Or perhaps the kind heading to Mars with the 2020 rover.

Considering that 95%–98% of Earth's life is underground, probing far below the surface, beyond the reach of radiation, might just give us the answer to one of the oldest age-old questions: Are we alone?

Of course, answering that question leads to others. For a scientist, the first question would ask whether or not it's actually alien life. Asteroids and comets smacking into Mars and Earth have sent chunks from each flying into space and landing on the other for billions of years. Swapping spit across the cosmos, so to speak.

"I think we've been exchanging microorganisms all this time," Gil Levin says. "And while they may have developed differences because of their respective environments, which are somewhat different—but not radically different—they're still much the same."

Those tiny interplanetary hitchhikers could've survived the long harsh trip through space in both directions. But to find out if Martian bugs are truly alien, we need to find living creatures and compare their genetic chemistry to that of life on Earth. If they're truly different, we'll know for certain that they're not our very, very distant cousins.

"In one solar system, there would be two planets that would've evolved life independently, distinct from one another," Lee says. "Therefore, life has to be some sort of automatic process for evolved chemistry, and therefore life must be common in a lot of places in the universe."

Alien life would prove we're not alone. So what would that say about humankind? Just as Copernicus and Galileo's radical ideas about the organization of the solar system forced us to rethink our place in it, how might life outside Earth throw religious belief systems into a tizzy? Are we as special as we thought? Or just another biological blob that God scattered throughout the solar system? For some, it would be a thrilling thought. For others, it might completely ruin their concept of heaven, hell, and the ways in which they lead their lives.

On the other hand, if DNA sequencing were to show a similarity and demonstrate a relationship between life on both planets, then the question becomes, how common is life? Did it spread anywhere else?

The search will continue. But just because NASA hasn't yet found Martians doesn't mean that others haven't. People have been encountering them for centuries. It's just a whole lot easier when you get rid of all that science stuff.

Launch of the *Opportunity* rover, July 7, 2003.

CHAPTER 3
CLOSE ENCOUNTERS
OF THE MARTIAN KIND

If you've spent a weekend binge-watching *Ancient Aliens*, you might think extraterrestrial visits go back thousands of years. Some think they date back to the Bible and that descriptions of Jacob's ladder to Heaven and Ezekiel's vision of a flying chariot were UFO sightings, not wondrous, divine actions. Once you buy into that, you might as well accept that aliens built the pyramids, because how else could weak humans with crude tools create such magnificent structures? These theories are entertaining to think about and make for good television, but there are no surviving documents that explicitly describe alien encounters. Theories are just theories. So, let's fast-forward through biblical times and ancient eras to the Age of Enlightenment when intellectual, scientific, and philosophical thinking swept through Europe. And Emanuel Swedenborg visited the spirits of Martians.

SPIRITUALISTS DON'T NEED SPACESHIPS

Before his Martian encounters, Swedish inventor and scientist Emanuel Swedenborg (*below*) dabbled strictly in Earthly matters, like engineering, geometry, chemistry, metallurgy, anatomy, and physiology. Swedenborg's work helped advance Sweden's mining industry and offered ideas about nerve cells and the structure of the brain that were well ahead of their time. By his fifties he had mastered the natural sciences and the physical world and decided to pivot to the spiritual side. His book *Heaven and Its Wonders and Hell from Things Heard and Seen* details his out-of-body experiences and the things he learned from angels, all of whom, he said, were once people on Earth.

But Swedenborg's travels didn't stop there. The universe offered so much more to explore, and he didn't want to miss any of it. Fortunately for the rest of us who haven't mastered the art of intergalactic astral projection, he recorded his adventures in the thoroughly titled *Earths in Our Solar System Which Are Called Planets, and Earths the Starry Heaven Their Inhabitants, and the Spirits and Angels There.* The spirits of Mars, in his opinion, are "among the best of all spirits who come from the earths of this solar system." One of the Martian spirits approached Swedenborg at his left temple and allegedly whispered into his ear "like a very gentle breeze." In their conversation, he learned that the people of Mars live in peaceful civilizations with no system of government.

"They converse with each other on what passes in their societies, and especially in heaven, for many of them have open communication with the angels of heaven," Swedenborg wrote. Any Martians who disrupt their happy communities are banished "and left to themselves alone, in consequence of which they drag on a most wretched life, out of society, among rocks or other places, for the rest no longer trouble about them."

Physically, the Martians were similar to humans, except the upper part of their faces were yellow and the lower part black, like Homer Simpson with darker stubble. Vegan Homers, to be more specific. They mainly ate fruit and stitched their clothing from the fibers of trees. They were also a pious people who, according to Swedenborg, "acknowledge and adore our Lord, and that He governs both heaven and the universe."

Based on his interactions, the Martians seem like nice folks, not like the "savage and almost brutal" Venusians he met, and not the invasive, destroy-Earth types that would later be described by science-fiction writers.

Swedenborg's concept of a life after death coincided with the rise of Spiritualism in the mid nineteenth century. In 1848 two young girls, Margaret and Kate Fox, claimed to hear ghosts rapping on tables in their Hydesville, New York, home. The excitement gave birth to a worldwide craze of séances, crooked mediums, and a belief that the dead could talk—despite that decades later the girls admitted to making the sounds with their toes. By the time they made their confession, it was too late. People wanted to believe their loved ones were still with them, especially after the massive casualties of the Civil War, and there was no shortage of clever mediums ready to indulge them. At séances, people frequently experienced tables lifting off the ground, objects floating in the room, bells mysteriously ringing, messages allegedly written by ghosts; sometimes they even heard the comforting voices of the dead speaking directly to them. These Spiritualists were spectacular magicians except, unlike stage entertainers, they claimed their powers were real and preyed on the bereaved. The dead were very profitable.

So, as the nineteenth century progressed, people believed in ghosts and people believed in Martians. And when the two mix, you get a whole new niche market for creative, opportunistic mediums. Enter Henry A. Gaston.

In 1880 Gaston, a lawyer in Chicago, claimed to have followed in Swedenborg's spiritual footsteps, but with a much longer stay on the Red Planet. Writing under the name "A Spirit Yet in the Flesh," he detailed his adventure in *Mars Revealed; Or, Seven Days in the Spirit World*. As his story explains, Gaston was relaxing under an oak tree one fine spring day when a "man of most magic form" and a "forehead that beamed with intelligence" approached and asked why he was sitting around doing nothing when he could be out visiting the planets and meeting God. Like anyone else spending their afternoon on Earth, Gaston explained that he didn't know he could do that. Now that he did, he suggested a trip to Mars. All he had to do was leave his body behind and let his spirit take a quick vacation.

One mystical flight through space later, Gaston and his companion arrived and found themselves surrounded by "beautiful valleys nestled at the feet of mountain peaks, filled with quaint homes, and surrounded by delightful parks and lawns." The many trees, animals,

MARS REVEALED;

OR,

SEVEN DAYS IN THE SPIRIT WORLD:

CONTAINING AN ACCOUNT OF THE

SPIRIT'S TRIP TO MARS,

AND HIS RETURN TO EARTH;

WHAT HE SAW AND HEARD ON MARS, ETC.

WITH VIVID AND THRILLING DESCRIPTIONS OF ITS

MAJESTIC SCENERY; ITS MOUNTAINS; ITS VALLEYS, RIVERS, LAKES, AND SEAS; ITS PEOPLE, TEMPLES OF LEARNING, WORSHIP, RELIGION, MUSIC, MANNERS, CUSTOMS, LAWS; ITS HIGHLY CULTIVATED AND PRODUCTIVE LANDS; TOGETHER WITH ITS BEAUTIFUL PARKS, AND ITS DELIGHTFUL PARADISE.

Being a work full of diamonds of thought, and of absorbing interest: A thrilling poem, in beautiful prose.

BY

A SPIRIT YET IN THE FLESH.

SAN FRANCISCO:
PUBLISHED FOR THE WRITER, BY A. L. BANCROFT & CO.,
721 Market Street.
1880.

"I was told that no snakes, rodents, or other noxious reptiles or vermin annoy the husbandmen in the field; for Mars' people have not cursed their planet by their sins, and therefore are not so punished as are the men of Earth."

—Henry A. Gaston
aka "A Spirit Yet in the Flesh,"
singing the praises of Mars in *Mars Revealed*

birds, large cities, and orchards filled with fruits, vegetables, and grains made the place seem like paradise. Things got even better when Gaston met the Martians, who were apparently no different in form than Earthlings—except for the women, who dazzled him with their flawless features, perfect teeth, brilliant minds, and graceful moves. He'd never seen their equal at home.

Gaston found these beautiful and busy folks to be as religious as Swedenborg had said. In fact, they were so devout that nearly all their conversation revolved around the discussion of God and his laws and powers. Gaston joined a Martian family at dinner and, having been given the power to understand the Martian language, heard all about God's awesomeness as each member of the household took a turn singing the Lord's praises. Their advanced telescopes even made it possible to see the Almighty ruling from his "central planet" filled with "myriads of people, who were rejoicing in his presence, and delighting in his goodness, his mercy, and beneficence." God seemed to be living it up on his very own celestial orb. As for their temples, the Martians didn't bother with stone or marble; they went straight for ultra-extravagance by using nothing less than gold, silver, and diamonds. Most things, in fact, were built with what we humans would call precious metals and stones—even the underground pipes that distributed water.

> ## "WHO CAN DESCRIBE A DAUGHTER OF MARS? THE PEN SHRINKS FROM THE DELIGHTFUL, BUT IMPOSSIBLE TASK."
>
> **—Henry A. Gaston,**
> aka "A Spirit Yet in the Flesh,"
> smitten by Martian women in
> *Mars Revealed*

"From this and other things you have seen you discern the energies, and wisdom, and genius of the people of Mars," Gaston's spirit guide proudly noted. These geniuses were so advanced that they discovered gravity 91,000 years ago and electricity 70,000 years ago. Beyond science, they understood life and its eternal nature. Death on Mars was simply a transition into immortality in the spirit world.

Poor Gaston must have felt like a fool—Isaac Newton had only figured out gravity centuries earlier and Thomas Edison invented the electric bulb just a year before *Mars Revealed* was published. Slowly but surely, we silly Earthlings were getting there. Mars would help our cause, and Gaston hoped his book would show the world that the more we caught up with the superior Martians, the more we'd understand that Spiritualism was a true religion and the next natural evolution of human knowledge. His imaginative descriptions of the Red Planet and the perfectly peaceful Martian society preached the importance of following God's laws, doing his will, and treating people with respect and goodness.

"If these simple precepts were obeyed by all on Earth as they are obeyed by all on Mars," he wrote, "Earth would become, like Mars, a paradise, and all Earth's people would become fit associates for the pure spirits who obey the commands of God." Who needs Heaven on Earth when you can have Mars on Earth?

Of course, like any good business-minded medium, and in the long and sleazy tradition of preachers trying to sell religion, Gaston was looking to make a buck. By fusing Spiritualism with Martian madness, his book was poised to do just that. His parting words were to remind readers to "encourage all you can, to buy and read a book so filled with good and interesting facts and truths, and with descriptions of what I saw and heard on Mars; for, if you do this, Heaven's rich blessings will attend your path and the pathway of all those who read it as their own; and all who act upon the fundamental law which rules all Mars."

There's always a catch.

Hélène Smith never claimed to have read *Mars Revealed*, but in the mid-1890s this Swiss medium jumped on the Martian bandwagon and started visiting the Red Planet, too. Smith, whose real name was Catherine-Elise Muller, began her mediumship in 1892 and gained notoriety for having channeled the spirit of Victor Hugo in her séances. The French author had died in 1885 but allegedly remained prolific in the afterlife by writing verses for Smith and her friends. Sadly for any Francophiles in the spirit circle, Hugo's stay was short-lived. Smith was soon overtaken by what were believed to be the spirits of her previous existences, like Marie Antoinette and her former lover Joseph Balsamo, known as an occultist and magician. The couple spoke and wrote letters to each other through Smith. At other times, the medium hosted the spirit of fourteenth-century Princess Simandini of India. Unusual as these spirit manifestations were, they didn't compare with her somnambulistic spiritual voyages to Mars.

Psychologist Theodore Flournoy began observing Smith in 1894 and, not surprisingly, his notes of Smith's Martian visions were quite different from Gaston's. Flournoy had hope for Smith, though, since the idea of a spirit traveling to Mars and discovering extraterrestrial civilizations appealed to him. Perhaps she could do what science couldn't.

"As soon as this vague hope takes shape," he wrote, "nothing seems to prevent its immediate realization; and the only cause for wonder is found in the fact that no privileged medium has yet arisen to have the glory, unique in the world, of being the first intermediary between ourselves and the human inhabitants of other planets; for spiritism takes no more account of the barrier of space than of time."

Once in a trance, Smith offered vivid descriptions of what she witnessed, like carriages that glided by without horses or wheels, houses with fountains on their roofs, and at least one very strange two-foot-long beast sporting a flat tail, the "head of a cabbage" with a big, green eye in the middle, six paws, and many ears. She also encountered big snails which, if they were especially big, may have been creepier than the cabbage-headed beast. As for the Martians, Smith said they were just like us, except that both sexes wore baggy pants and long blouses drawn tight at the waist—a description that may have been influenced by her countryside neighbors.

While Gaston claimed to automatically understand Martian, Smith convinced people that she both spoke and wrote the language fluently. Flournoy described séances in which Smith's tongue seamlessly shifted

A Martian landscape, by Hélène Smith.

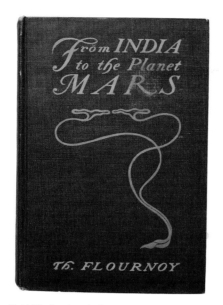

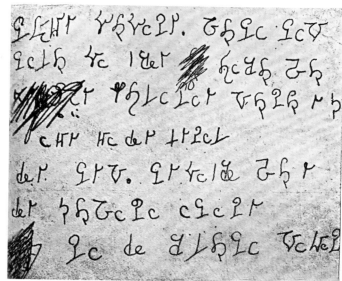

Right: Martian handwriting, as written during a séance by Hélène Smith on October 10, 1897. From Theodore Flournoy's *From India to the Planet Mars*.

to "Martian" as she continued to "chat with us in the most natural manner." No one understood a word of it.

Over the course of several years, Flournoy studied Smith's spirit personalities and the structure of the Martian language in enough detail to fill a book—which he did. In *From India to the Planet Mars*, he concluded that Smith suffered from a multiple-personality disorder and her imaginative descriptions of Mars were a subconscious reflection of the terrestrial world. As for the Martian letters and vocabulary, he broke it down and determined it was nothing more than "French metamorphosed and carried to a higher diapason." At times this included random sounds with no meaning, "analogous to the gibberish which children use sometimes in their games of 'pretending' to speak Chinese or Indian."

Crazy as it all sounds today, Smith had her believers. "Her character and standing are strongly vouched for," one newspaper stated in its review of Flournoy's book. "The phenomena which are minutely related in this volume are such as pertain to the clairvoyant or trance state, and are of an extraordinary character."

With Spiritualism booming all over the world, it's no surprise that another medium found her way to Mars. Mrs. Smead was an American psychic who had dabbled with planchette writing as a child but didn't discover her full powers until 1895—just as Smith was having her visions. Six years later, Smead's husband, a clergyman, contacted psychologist and psychic researcher James Hyslop to investigate his wife's strange and fantastic abilities. Her revelations about the Red Planet came through the spirits of her three deceased children and brother-in-law, Sylvester. Each guided Smead's planchette to write or draw information about Mars, where they were spending their afterlife.[9]

---

9 FYI, Smead's family also mentioned that Jupiter is "babies' heaven."

Hyslop initially approached the case with caution, knowing there was no shortage of fraudulent mediums looking to deceive anyone and everyone. But once he got to know the Smeads, he found them to be "honest and conscientious people" and may have let his guard down. Comfortable and quite curious, he began recording his experiences and observations of their alleged communications. One of the early conversations with their daughter, Maude, went like this:

Mr. Smead: Do the people in Mars have flesh and blood as we do?

Maude: Yes.

Mr. Smead: Do they look like us?

Maude: Some.

Mr. Smead: Are there big cities there?

Maude: No. The inhabitants are most like Indians.

Mr. Smead: American Indians?

Maude: Yes.

Mr. Smead: Are they highly civilized?

Maude: Yes, some are, in some things.

Mr. Smead: What things?

Maude: In fixing the water.

Mr. Smead: How in that way?

Maude: Making it so that it is easy to get around it.

Mr. Smead: How do they do that?

Maude: They cut great canals from ocean to ocean and great bodies of water.

This conversation took place shortly after Percival Lowell published an article about the canals on Mars. Smead claimed she hadn't read it. Yet it's the most detailed answer in a rather dull exchange about life among extraterrestrial beings. Maybe if Mr. Smead spoke more enthusiastically to his dead daughter living on Mars, he could have elicited more interesting and descriptive answers.

Still, subsequent communications revealed many other details about Mars and its inhabitants. For example, the men who resembled American Indians wore dresses and pants and kept long hair tucked beneath their hats, whereas women kept their long hair hanging under their "funny hats" (*opposite*). These very Earth-like people built their houses on the shores of lakes or canals and spent time creating beautiful embroideries. Lowell would have expected them to live near their water source and vegetation, but the artful needlework would have surely been a pleasant surprise. These crafty Martians were equally inventive, too. Air ships of "very peculiar and ingenious construction" were powered by coils that needed constant winding to flap their wings. The design sounds crude, but the men of Mars who could dig canals like nobody's business also appeared to have us beat in the aeronautics department.

MAN.   WOMAN.

Smead's planchette even sketched out Martian hieroglyphs, which they translated into English letters. For example, *"Moken irin trinen minin aru ti maren inine tine"* meant "Flowers bloom there. Many of the great men plant them." (No explanation was given for how they made their translations.)

If life on another planet seems like it should have more spectacular differences, Hyslop agreed, noting that the similarities to terrestrial phenomena were suspicious and "take the Martian 'communications' entirely out of the category of spiritistic revelations." Instead, the psychologist suggested Mrs. Smead's revelations were the result of a secondary personality. Flournoy closely followed Hyslop's studies and added, "the Martian revelations of Mrs. Smead present the same character of puerility and naive imagination as those of Mlle. Smith—although they differ greatly in details—I cannot but think that the psychological explanation is at basis the same."

Despite their evaluations and the descriptive discrepancies, some still clung to Spiritualist beliefs and argued that the Mars revelations could be genuine.

"The two mediums may have been seeing different parts of the planet," one optimistic reporter wrote. "A Martian medium studying Timbuctoo would get an entirely different impression from one studying Boston."

Fair point, but it exemplifies the desperation of people to make extraterrestrial contact. Suddenly, they were rationalizing mediums and ridiculing science. Why should we waste money on giant mirrors or make Professor Todd float in a balloon waiting to hear from Martians when we could just communicate through psychics?

"The Martians, if they are as advanced in their development as they are supposed to be, have doubtless long since abandoned such crude methods of communication as letters and electric vibrations and are using thought-waves," suggested one newspaper.

Dr. Hugh Mansfield Robinson, a London lawyer, agreed. In 1926 he reported that he was having a telepathic affair with a tall, big-eared Martian woman named Oomaruru.

PSYCHIC NEW YORKERS
AND HAIRY MARTIANS

Sackville G. Leyson, president of the Society for Psychical Research, proved the worthiness of his title by experiencing his own psychical journey to Mars. It only took his spirit forty minutes to make the trip. His body stayed home in Syracuse, New York.

Upon Leyson's arrival to the "big globe of fire," he encountered two species of Martians. "One so large I only came up to their knees, and one so small they only came up to my knees," he described. Both were covered in hair and wore no clothes. "The largest species had huge ears, a nose like a lion, and only one eye, in the middle of the forehead, … the small ones have two eyes, one in each temple. They had no noses, but there was a hole in each cheek," he added. These dwarfish Martians also had webbed feet and could walk up walls.

Leyson noted that the two breeds lived apart from each other. The little fellows lived in holes in the ground or rocks, whereas the big cycloptic race lived in houses made of rocks. He failed to report how this Martian civilization functioned, but he did spot a few working with machinery that guided lights across transparent rocks, which created rays visible from Earth. Perhaps he believed these lights were what Francis Galton had seen a decade earlier.

Shortly after Leyson's bold claims, another New Yorker, Pauline Corri, came forth to describe an encounter with a similar one-eyed hairy Martian six years prior.

MAKES TRIP TO MARS, SEES RACE OF CYCLOPS

This Is Wild Story Told by Man Whose Spirit Briefly Left Body.

SYRACUSE, N. Y., Aug. 19.—Sackville G. Leyson of this city recently paid a visit to Mars, and although the distance is 111,000,000 miles, he went there and back in forty minutes—at least his spirit did, while his body was in his residence, 131 South avenue. Mr. Leyson is president of the Society for Psychical Research. In describing his visit he said:

"When I approached Mars it looked like a big globe of fire, and it seemed as if I was about to plunge into a molten mass. It was surrounded with blood red clouds, mixed with others of a greenish hue.

"There are two tribes of people on Mars—one so large that I only came up to their knees, and the other so small that they only came up to my knees. None wore clothing and all were covered with hair.

"The larger species had huge ears, a

MYSTIC'S SPIRIT SAW MARVELOUS SIGHTS IN MARS

Leyson, in Trance, Beheld Men Who Looked Like Cyclops.

BODIES HAIR-COVERED

Their Noses Are Like Those of Lions and Their Lungs Are Crosswise.

ROCKS ARE TRANSPARENT

Little Web-Footed Men Walk on Walls Like Flies.

She didn't have to spend forty minutes traveling spiritually to Mars, though. This Martian made his own trip, appearing directly in her home.

"That my visitor was not flesh and blood I knew, and while I waited, with my flesh creeping, I heard from somewhere, but not the man, 'He is a spirit from the planet Mars,'" she recounted. Then he vanished, never to make anyone's flesh creep again—at least, not on Earth.

Leyson planned to make another trip, with scientists and psychologists present to verify the truth of his tale, but no reports confirm that the psychic traveler's spirit ever made a return.

VISITS PLANET OF MARS; FINDS ONE-EYED GIANTS

Syracuse (N. Y.) Man Brings Back Queer Account of Our Nearest Terrestrial Neighbor.

Syracuse, N. Y.—Sackville G. Leyson, president of the Society for Psychical Research, says he recently paid a visit to Mars. Although the distance is 141,000,000 miles his spirit went there and back in 40 minutes

GEE! HOW FUNNY THEY ARE

LOOK AT THE LOOKS OF THAT WILL YOU

MARS

Leyson Interviewing the Martians.

CTQ

OCT 27 26   32760

The Secretary,

I have to report that Dr. Mansfield Robinson of Spital Square called here yesterday to ask if we would transmit a message to the planet Mars at a fixed time from the Rugby Station. I consulted Mr. Phillips as to the tariff to be charged which was 1/6 s.d. per word. In view of the nature of the message and the importance which Dr. Mansfield attached to it, suggested a dual transmission. Dr. Mansfield Robinson accepted the arrangement and has paid the money for the transmission of three words at 11.55 tonight.

He asked us what wavelength would be used from Rugby and we informed him. He said that he had other means of communicating with the planet and he wished them to know the exact wavelength and the time at which the message would be transmitted. He asked us could we possibly put a receiving station at his disposal at the same terms to receive messages from Mars, and we asked him what wavelength, and he said the wavelength used by Mars was 30,000 metres. We told him that it was not possible to give him a receiving station, but we suggested other means by which he could get the information.

Dr. Mansfield Robinson is singularly serious in this business, and we may expect other messages from him. I do not think we could have any conscience pricks as regards taking the money for he is perfectly sane and seems to have devoted his life to the study of possible intercommunication with the plane In fact he said that a colleague of his had obtained as long as four years ago definite signals from Mars, but that he was a "weak-kneed scientist" who refused to allow the information to be published. The message consisted of three words quite cryptic to us.

John Lee

"THEY LAUGH AT OUR SCIENTISTS."

If you asked Robinson, all those scientists who speculated that Martians would be tall were spot-on. His intergalactic girlfriend told him so. Oomaruru stood over six feet and had dark penetrating eyes, a curious nose, and a half-smiling mouth. The men reached heights of more than seven feet. They were gentle giants who only wanted peace. In many ways, they were much like us: living in houses, driving cars, and enjoying simple pleasures such as drinking tea and smoking pipes. But when it came to food, they were far more advanced. Martians electrified their fruit trees, which gave the produce all the nutrients these life-forms needed. Except perhaps the lower forms of life, which were beings described only as having heads shaped like walruses.

New to this whole world of interplanetary telepathy, Robinson wrote to psychic researcher Harry Price for assistance. In his letter, he claimed to have patented an instrument called a Psychomotormeter that allowed him to communicate with Oomaruru and her peers. He heard their voices through a floating trumpet and a medium and wanted Price to record them on a Dictaphone.

Price agreed, and Robinson set up a Martian séance with his medium of choice, Mrs. St. John James. Oomaruru would be waiting to receive their mental missives.

"She is very pleased at the idea of being treated with scientific seriousness," the doctor wrote Price. He added that "a very cultured giant" called Pawleenoos would participate as well.

When the day arrived and the session began, Mrs. James fell into her trance and channeled numerous messages from the Martians. Controlling the medium's pen, Oomaruru wrote the full Martian alphabet, which did not resemble Smith's or Smead's, but did look like a series of scribbles from someone trying to write with their nondominant hand. Pawleenoos sketched his portrait, and a Martian princess made a special guest appearance to sing a love song.

"If the song was not Martian, it was certainly 'unearthly,' and sounded rather like a solo by a crowing cock," Price recalled. "There was nothing musical about it."

By fall of 1926, despite Robinson's self-proclaimed telepathic abilities and his apparent success with his medium, he turned to radio waves at the post office for further communication attempts. Mars was then in its closest position to Earth—just thirty-five million miles away—so the doctor thought the signals might be able to reach it.

A memo dated October 27 from London's Central Telegraph Office (*opposite*) explains that Robinson arranged a transmission to Mars for its standard long-distance rate of just eighteen pence per word (about thirty-five cents), and that he wished to know the exact wavelength and timing. Apparently, Oomaruru could share this information with the director of the largest wireless station on Mars.

"Dr. Mansfield Robinson is singularly serious in this business and we may expect other messages from him," the memo reads. "I do not think we could have any conscience pricks as regards taking the money for he is perfectly sane and seems to have devoted his life to the study of possible intercommunication with the planet."

The message was scheduled for 11:55 that evening using the doctor's requested wave-

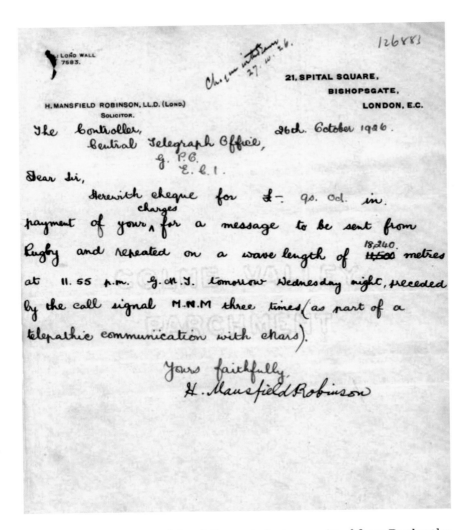

length of 18,240 meters (*above*). It was to be transmitted from Rugby, the largest and most powerful wireless station in the world at the time, and consisted of three cryptic words: Opesti, Nipitia, Secomba.

Postal clerks stood by with a receiver tuned to a wavelength of 30,000 meters, which Robinson said was the Martian's preference. Sadly, they received no response. But Robinson got a telepathic heads-up from Oomaruru that claimed she received only a "repetition of the letter M."

Even people who advocated Spiritualism found the event a bit far-fetched. British physicist Sir Oliver Lodge, for example, was quoted as saying, "We have not got into touch with Mars, and I doubt if we are likely to do so. If we did get anything there, how are they going to know what we are talking about? They do not understand the Morse code; neither do they understand the English language. So how are they going to understand?"

Robinson didn't trouble himself with such petty concerns. Instead, he waited two years for the Red Planet's next opposition, and then he made another attempt (*below*).

Once more, the post office agreed to dispatch his message. A memo stated that its reasons were "mainly in order to obtain free publicity for Rugby." After all, press coverage would promote their long-distance rate far more effectively and efficiently than paid advertising, officials reasoned. The memo further noted that "it was made clear that we did not regard ourselves as co-operating in a scientific experiment, but were merely performing the service at his request as a commercial proposition."

This time, editors from two London newspapers wrote to the post office requesting a presence at the transmission, but the Postmaster General rejected them. The message, which read, *"Mar la oi de Earth,"*

A G E N D A (1).

Inter-planetary wireless test for Dr. H. Mansfield Robinson
of 21, Spital Square, Bishopsgate, E.1. (Telephone:
Bishopsgate 7583).

Transmission required

    On 18,750 metres with fullest power available
    A.M
    a. At 2.15 G.M.T. in the early morning of
Wednesday, 24th. October in the direction of
the planet Mars about 60° above horizon in S.E.

      "M M Loy to Mars bx Erth"
      A.M.
    b. At 2.30 G.M.T. same morning.

      "M M God is loy"

No repetition or alteration of the above signals
should be made. The exact phonetic spelling
above must be used. X *is one letter meaning from*

Owing to over-lapping of signals through interference
across this vast distance the message must be sent
at about six words per minute, the dashes about three
seconds long and the dots about 1 second long
with intervals of three seconds between each and six
seconds between each word. As a new means of determining
the distance between the planets the exact time of the
beginning and end of each message should be recorded.

SCIENTIST HOPES BIG-EARED MARTIANS YET MAY GET RADIO

Admits He Has Received No Reply From Planet to His Message.

Insists, However, Friends Above Are Anxious to Exchange Good Wishes.

LONDON, October 25 (*P*).—Small ears and long antennae in England strained in vain to catch a return message from the big-eared folk of Mars, to whom a wireless message was dispatched yesterday morning.

Dr. Mansfield Robinson, author of the message, who professes acquaintance with the Martians through telepathic means, clings stoutly to his faith in the possibility of interplanetary conversations. He admits that no response was received, but insists that his friends up yonder are anxious to exchange good wishes with him.

The wording of the message to Mars remains a mystery. The postal authorities collected their 1 shilling 6 pence per word. It is purely a matter of business to them, and they are not at liberty to disclose the contents of the radiogram.

Wants Greater Wave Length.

This went on an 18,700-meter wave length, and not a sound came back. Dr. Robinson thinks the wave length

MISS MARS OF 1928.
Sculptor's conception of "Oomaruru." Beauty of the distant planet Mars, with whom Dr. Robinson of England says he communicates by telepathy.

translated by Robinson as "Love to Mars from Earth") was transmitted at 2:15 a.m. on October 24. Even though the media was denied attendance, the press gave it a great deal of coverage. One paper even reached out to a professor at the Yale Observatory, who believed radio signals to Mars were futile and that "the inhabitants of Mars may be more hideous than we can imagine—with no heads, perhaps, and maybe no brains."

The professor was proven correct on the "futile" part. Again, no response came through.

All was not lost though, for Oomaruru continued to chat with Robinson telepathically. "Go home to bed, but do not be downhearted," she told him after the failed attempt.

Robinson, however, blamed the post office's inept equipment. "The wavelength of 18,700 meters used by the post office does not go through the heavy-side layer of rarefied air, and therefore signals are reflected around the Earth," he told the press.

"The Martians were very annoyed that the signals could not come to them," he added. "They were sitting up for hours to receive signals. They laugh at our scientists because they themselves have got rid of atmospheric troubles altogether, and yet we have not."

Undeterred, Robinson decided to try another transmission in December, but this time from Brazil, using 500-kilowatt signals with wavelengths at more than 21,000 meters. Similar to his previous message, this transmission read, "God is love; from Earth to Mars."

Unfortunately, changing hemispheres did not change the results. A Brazilian newspaper facetiously suggested that the Martians probably didn't understand English, and that Robinson should have tried sending his message in Portuguese or another language, "on account of the greater difficulty of Shakespeare's tongue."

Robinson's mission went quiet until January 1930, when he apparently dispensed with radio communication and reported that he'd telepathically conducted an interview with Oomaruru for the United Press. In his alleged conversation, she suggested opening a College of Telepathy. She also expressed dismay at the lack of world peace after nearly two thousand years of "listening to the teachings of Jesus."

Telepathy, Oomaruru believed, would make our world a better, more peaceful place. Not only would it help us reach Mars, but it would also help simplify communication right here on Earth. Robinson believed it was the "missing link in progress" and felt it had been neglected by scientists "in spite of its obviously tremendous value."

He was also frustrated by the inefficiencies of the telephone, which telepathy would solve. "One in three calls do not materialize," he said. "You either get a wrong number, a busy signal, or a 'they don't answer' from the operator. All this will be done away with when the world learns to telepathize."

A benefactor planned to help fund the college with the sale of artwork valued at nearly a quarter million dollars. This mysterious artist apparently made good on his promise, because six months later Robinson opened the College of Telepathy, staffed with six teachers and a telepathic dog named Nell. To help build the student body, he offered free tuition to the first seven pupils. Each would be required to practice "complete chastity and abstinence from flesh, alcohol and tobacco, coupled with good health and a desire to seek spiritual development." Nell's role was not clarified.

Only one student is known to have enrolled. Her name was Claire, and in addition to meeting the aforementioned requirements, she also signed a membership declaration in which she agreed to obey its rules, keep its secrets (unless "compelled by a competent court of justice"), and use her powers for the benefit of others. Robinson had hoped for more female students because he didn't believe men would be able to beautify their souls and bodies "even for the sake of telepathy."

No further details on attendance or successes were ever reported. However, Robinson made headlines again in 1933 after claiming to be telepathically in touch with Cleopatra, who was living on Mars as a farmer's wife. This may not have pleased his own wife, who once told reporters that she wouldn't allow his experiments to be conducted at home. "There will be no more of that foolishness in this house," she said.

Telepathic and spiritual travels to Mars may have been the thing to do in the first half of the twentieth century, but finding Martians right here on Earth was all the rage of the second half.

FLYING SAUCERS?! BLAME MARS.

June 24, 1947, was the day the UFOs began their alleged domination of the skies and invasion of the world's imagination. It was on that day, at 3 p.m., that a pilot from Boise, Idaho, named Kenneth Arnold saw nine shiny saucer-like objects flying over Mt. Rainer in western Washington.

"I could see their outline quite plainly against the snow as they approached the mountain," Arnold told reporters. "They flew very close to the mountain top . . . like geese in a diagonal chain-like line, as if they were linked together. . . . A chain of saucer-like things at least five miles long, swerving in and out of the high mountain peaks."

He claimed they were flying at speeds of up to 1,200 miles per hour.

"Everybody says I'm nuts," Arnold told reporters after sharing his story.

Nuts or not, the government probably wished the aliens abducted him and whizzed off into space without a trace, because Arnold's account set off a massive wave of similar reports all across the country. That drove the Army and Air Force even more nuts. In the weeks that followed, people in thirty-three states claimed to have seen similar disk-shaped objects zipping through the skies, including saucers with spots, fiery tails, American flags, and some that broke in half and made strange noises. One woman even saw something she described as looking more like washtubs "the size of a five-story house."

An Army spokesman claimed no high-speed experiments were being conducted and seemed as baffled as everyone, saying, "As far as we know, nothing flies that fast except the V-2 rocket which travels at about 3,500 mph and that's too fast to be seen." But a commanding officer of the Army's rocket proving grounds at White Sands, New Mexico, said the alleged saucers must have been jet airplanes.

The public wasn't buying it. Nor did it believe the sightings were balloons, optical illusions, solar reflections, or radio-controlled missiles developed by the military. A Soviet foreign minister visiting the United Nations wryly suggested that the British might have been exporting too much Scotch whiskey to America, or that the saucers may have been due to a Russian discus thrower "training for the Olympics Games who does not realize his own strength."

Orville Wright, one half of the team that made flight possible for us humans, weighed in as well. He didn't buy into any of the alien speculation. Instead, Wright thought the U.S. government was prepping the country for another war by creating propaganda "to stir up the people and excite them to believe a foreign power has designs on this nation."

> "I DON'T TAKE MUCH STOCK IN THE IDEA THAT THEY'RE MEN FROM MARS BECAUSE THEY'VE BEEN AROUND FOR ALMOST A MONTH. A MAN FROM MARS WOULDN'T BE AROUND THAT LONG. ONE LOOK AT WHAT GOES ON HERE AND HE'D HEAD RIGHT BACK FOR HIS PEACEFUL PLANET."
>
> —**Gracie Allen,**
> *Louisville Times* columnist, commenting on the numerous flying saucer reports in 1947

A Chattanooga, Tennessee, watchmaker may have agreed with the Wright brother when he accused the War Department of stealing his idea for a "flying saucer" machine powered by a rubber band. He claimed to have submitted his invention to officials in 1943 but was rejected because it wasn't practical "at the present time." The amateur inventor figured that times had changed and now they'd ripped him off, replacing the rubber bands with atomic power.

The flurry of sightings was merely a warm-up act for the main event: Roswell. Barely a week after Arnold's sighting, a New Mexico rancher spotted some shiny objects on his land. He collected all the curious scraps and brought them to the attention of local law enforcement.

"I asked the sheriff to keep it kinda quiet," he said days later. "I was a little bit ashamed to mention it, because I didn't know what it was."

Unfortunately, a press agent for the Army base issued a statement announcing that officials had found a "flying disk." With those two words, the U.S. government admitted the existence of UFOs, and Washington, like the rest of the country, began freaking out. These flying saucers were *real*? It didn't take long for rumors to spread that officials had carted away the remains of extraterrestrial travelers. More specifically, they were small, humanoid figures with big heads and large oval eyes but lacked noses. (This simple Roswellian alien image has become so stereotypical it's even an emoji on your phone.)

The Air Force quickly retracted its "flying disk" statement and denied any possession of aliens. The alleged UFO, the public was informed, was merely a weather balloon. Years later it was upgraded to a nuclear test surveillance balloon connected to a top-secret scheme called Project Mogul.

As for the many, many other unidentified flying objects whirling about and unnerving Americans from coast to coast in the years following Roswell, well, the federal government was just as curious. Officials weren't really worried that aliens might be casing the United States, but with the Cold War under way they had to explore the possibility of Soviets invading U.S. air space with new technology. Two intelligence officers flew to Washington to meet with Arnold and inspect the area of his sighting. An anonymous caller told reporters that he'd seen "closely guarded fragments of a flying disk" loaded onto the officers' B-25 bomber. Tragically, that plane crashed during its return flight, killing both men. The Air Force wasn't spooked by the accident, though. Its studies of the copious claims led to its conclusion that the UFO sightings were a result of mass hysteria and hallucination, hoaxes, or simple confusion. It scaled back the visibility of its investigation, having realized that showing legitimate interest would only encourage more people to believe in flying saucers, leading to more sightings, fear, and a "war hysteria atmosphere."

Despite the Air Force's conclusions, it wasn't just the masses seeing UFOs. It was the military, too. That made things a little more con-

Right: A 1957 United States Air Force report detailed plans to develop its own flying saucer.

AVRO AIRCRAFT LIMITED

PROJECT 1794

1808-704-1

USAF PROJECT 1794

fusing. Like in 1949, when Commander Robert B. McLaughlin at White Sands became a staunch believer. The officer claimed his team spotted a UFO while monitoring a weather balloon they had set up fifty-seven miles northwest of the proving ground base. The Navy group and scientists were busily taking measurements when a flying saucer distracted them. McLaughlin described the strange disk as being just over a hundred feet in diameter, about sixty miles overhead and traveling at eighteen thousand miles an hour. That was Ludicrous Speed long before *Spaceballs*.

"I am convinced that it was a flying saucer and, further, that these disks are space ships from another planet, operated by animate, intelligent beings," McLaughlin wrote of the sighting.

That planet, naturally, was Mars. McLaughlin thought the nearby testing of the first atomic bomb in 1945 may have created a flash of light visible to any Martians watching us. The commander, like anyone who grew up in the early decades of the twentieth century, had surely read his share of headlines about intelligent Martians and our attempts to communicate with them. And like the many prominent scientists creating those headlines, McLaughlin believed that life developed earlier on Mars and gave its people a "big start in scientific development." Just as we were watching other planets, why wouldn't they be doing the same?

THE MARTIANS ARE HEADED
FOR HOLLYWOOD

All those Martians filling the skies with their flying saucers just wanted to be in the movies. Or at least that's what Carl Goerch, a local radio commentator and amateur aviator in Raleigh, North Carolina, told listeners of WPTF one unexpectedly bizarre and frightening summer evening in 1947.

Goerch's brush with Martians occurred during a short flight from Raleigh to Wilmington when a flying saucer swooped in and surrounded his plane. He knew it came from Mars because the license plate read: "Mars 86-4-77." Amazingly, they called their planet Mars just like we do. More remarkably, after billions of years of evolution, they, too, had developed English letters and Arabic numerals. These unbelievable details flew over the heads of listeners higher than the spaceships they believed in.

As for the whirling UFO, Goerch told his radio audience that "it looked something like the thing that discus throwers use, round but raised slightly in the center." When a panel opened on the craft, he got a good look at a group of five yellow-eyed, green-skinned Martians wearing silver helmets. One of them shouted at him through a megaphone. "I heard a strange and peculiar voice," he recalled. "It wasn't human nor did it sound like any animal I ever heard."

Somehow, he managed to interpret their Martian hollering to learn that they were completely lost and needed directions to Hollywood. Yes, they were advanced enough to get to Earth, but hadn't yet figured out Waze. By chance, Goerch had a large map handy and scribbled some directions on it, then held it up in the cockpit so the aliens could get a good look through the windshield. While they studied the map, he stealthily took photos. But darn it, his camera had no film. Of course, photos wouldn't have helped his radio listeners anyway.

Goerch's lack of proof—not to mention absurdity—didn't matter. With the flying saucer craze in full effect, he had unwittingly re-created the 1938 panic from Orson Welles's broadcast of *The War of the Worlds,* on a smaller scale. Calls from concerned Carolinians, Georgians, and Floridians flooded the station and jammed the WPTF switchboard for the next twelve hours. The control tower at Raleigh-Durham airport was barraged with callers, too, and disturbed the flow of airplane traffic.

The next day the imaginative broadcaster admitted he'd underestimated the gullibility of his audience.

THE RIDDLE OF THE
FLYING SAUCERS
is another World watching?

GERALD HEARD

Since no one on his team—or anyone on the planet—had been attacked, McLaughlin figured these curious visitors were just getting a closer look. "I can't feel that there is anything terrible, hostile or dangerous about the Flying Saucers or their occupants," he wrote. "So far, all I have suffered is a little hurt pride. They got here first."

McLaughlin offered no opinions on their possible appearance, but English author and philosopher Gerald Heard did. The Martians zipping around New Mexico weren't the twenty-foot-tall creatures with large chests and skinny legs that others had imagined. No, Heard's Martians were quite different. They were bees. A whole swarm of super-bees.

Heard had recently read an interview with University of Chicago astronomer Dr. Gerard P. Kuiper, who stated that the carbon dioxide and lack of oxygen in the Martian atmosphere meant that no life as we know it, except "a form of insect life," could exist. Heard ran with it. He was no entomologist, but he did know bees. In 1941 he had written *A Taste for Honey*, which featured a madman who trained an army of killer bees. After Kuiper's remark, Heard took his passion for the industrious, organized honey-makers and blended it with the UFO craze to write *Is Another World Watching? The Riddle of the Flying Saucer* (opposite). The book presents a thorough study of the bee and argues that they don't just act on instinct, but they also learn and evolve. As Heard imagined it:

> "It is pretty certain that they started far simpler in their way of life than they are now—just as we started without gear and goods and plants and tools and cities and transport. It is generally conceded that the bees, who now have cities and a hierarchical society, came up from solitary forms, many of which still survive. From that state they have built up cities and the organization of cities, control of population supply, distribution of power, and order of succession in a manner so masterly that beside it our own effort along these same lines still looks very amateurish and dangerously incompetent."

SUPER-BEES BELIEVED TO BE PILOTS OF FLYING SAUCERS

NEW YORK, April 10 (AP) — A new book speculates that pilots of flying saucers are super-bees from Mars, two inches long and quite beautiful.

The super-smart bees are Ger-Martian bees have stingers. They've existed on Mars so long that it's presumed they have no enemies, in a world "where intelligence has won total freedom from brutal, repressive force,"

So, Mars was filled with clever bees who make us mere mortals look like a bunch of knuckleheads. Plus, with their lesser gravitational pull, Martian bees—and their stingers—would grow larger. Now, if the UFO described by McLaughlin sounds like a feat of engineering outside the scope of a bee's abilities, then you're not giving these clever bees the credit Heard thought they deserved. Heard informed his readers:

> *"When we think of whether a space-ship could be built by a super-bee, we must remember, first, they are immensely ahead of the bees—or any insect—here. And when we study the hive, even at the level we know, it is a pretty amazing piece of accurate skill and real engineering. Now they have speech, they may soon turn to use other materials than wax for the hive structure, though the plan may be so good that it will not be changed much."*

Like McLaughlin, Heard thought the atomic bomb may have alarmed the Martians, who considered it a threat to their well-being. Not because the bees thought we'd blow them up, but because they feared we'd blow up the Earth, and its fragments would scatter and block what little sunlight they enjoyed. Fortunately for all of us, we haven't blown ourselves up and ruined any Martian summers.

Absurd as Heard's theory may sound, in March 2018, NASA announced a proposal to send a swarm of "Marsbees" to the Red Planet. These robotic bees, designed with wings the size of a cicada's, would provide scientists with a way to reach places the rovers can't. Armed with sensors, wireless communication devices, and rovers as a mobile base, the Marsbees could collect data and expand the possibilities of exploration. Add a touch of science fiction, and one can imagine a day well into the future when the Marsbees evolve and fly back to Earth, landing perhaps somewhere near New Mexico in a flying disk a hundred feet in diameter.

PRANKSTERS, SCREWBALLS, AND BELIEVERS

Cedric Allingham claimed to be the first man to meet a Martian right here on Earth.

The big event happened on February 18, 1954, while Allingham was strolling around the coast of a small fishing town in northern Scotland, just doing a little bird watching and enjoying being away from the hustle and bustle of London. But as he gazed to the sky to spot some feathered friends, he saw a flying saucer instead. So his story goes.

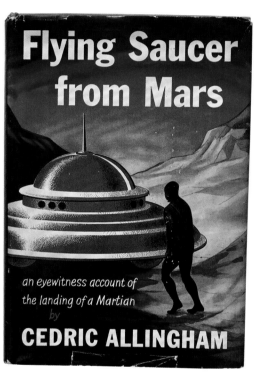

Luckily for him, the UFO landed nearby and gave him the opportunity to take a closer look. Allingham, who dabbled in astronomy in his spare time, estimated the ship to measure about fifty feet in diameter and about twenty feet high. "It was a magnificent craft, and its finish would surely be the envy of our aircraft manufacturers," he wrote in his 1955 book about the incident, *Flying Saucer from Mars*. "I could not say what the metal was; its color and luster were not unlike a polished aluminum (though it would of course need to be very much more robust)."

Based on photos published in the book, the ship looked like an old hubcap or a cereal bowl flipped over on a plate. A rudimentary drawing of an oval and a smaller semi-circle on top is labeled: "facsimile of rough memory sketch made by the author." This might lead one to think the book was authored by a five-year-old. But in fact, Allingham claimed to be thirty-two, the same age he estimated the Martian to be. His new alien friend was also similar in height, had the same haircut, and wore the same type of clothes. Physically speaking, the only difference Allingham noted was a higher forehead. Just when it sounded like he'd been staring at a distorted reflection of himself, he detailed one other minor difference. Being new to the planet, the Martian needed a breathing apparatus. "It seemed to be in the form of a tiny tube up each nostril, joined by a metallic band about as thick as a match," Allingham recalled.

The two species attempted a brief chat for about a half hour, but the language barrier made things tricky. "Our alphabet must be hopelessly inadequate to allow us to express their words phonetically," he said. Allingham sketched a diagram of the sun, Venus, Earth, and Mars to determine where the visitor was from. He also drew Mars with its canals and his friend nodded, confirming what so many had long suspected.

"So Lowell was right all along," he wrote. "Like so many pioneers, he suffered more than his share of criticism. The truth, however, nearly always emerges in the end."

The Martian had questions, too. After all, he didn't travel more than thirty million miles to point and nod at sketches. He wanted to know if there would be another war. Allingham shrugged his shoulders. How could he know?

Frustrated and disappointed, the Martian had enough of this place and walked back to his saucer. Only then did Allingham think to take a photo of what was arguably the most remarkable encounter in the course of human history. The image printed in the book shows the humanoid figure from behind, walking away. It could've been anyone with two arms, two legs, and a head.

For those inclined to believe, Allingham's arguments for the existence of Martians can be persuasive. One reviewer sounded ready to hop aboard the next UFO when he wrote, "As a book, Allingham's *Flying Saucers from Mars* is by a long, long way the best written, sanest, most unimpassioned and convincing that I have seen to date. . . . It has none of the occultism that is making other Saucer treatises ridiculous. It's the kind of story that could convince a jury."

Despite the vote of confidence, Allingham understood that he'd initially have doubters because Martians were something new that may take a little getting used to. Like McLaughlin, he suspected they finally made contact with Earth in those post-war years because our technology was improving and our weapons were becoming dangerous—to ourselves, and potentially to them. Allingham posits:

> *"Can you picture what would happen if we, in our present state of cultural backwardness, managed to establish ourselves on Venus or Mars? Wherever you find men of our kind, you find war; and we might well bring destruction to the whole civilization of our neighbor worlds, either by fighting them or by fighting among ourselves. We cannot blame those of greater intelligence for taking precautions."*

Martian or no Martian, Allingham makes a good point—which at least gives his Martian tale a purpose since it was revealed decades later that, not surprisingly, there was no Martian. In fact, there wasn't even an Allingham. The whole thing was a hoax constructed by a British astronomer named Patrick Moore, who had a penchant for practical jokes and apparently thought it'd be amusing to capitalize off the success of recent UFO publications.

Some twenty years after the launch of the flying saucer age, thousands of sightings continued to pour in, along with nearly as many explanations, not to mention UFO clubs and a library's worth of books that claimed to offer the "truth." If any of these alleged spacecraft were legitimate, they theoretically could have come from other solar systems, but Mars remained the prime launching-off point. After all, it's the only planet that anyone believed was capable of supporting life, even though most had accepted the canals didn't exist.

"I believe space ships are the only answer. The Earth has been under periodic observation from another planet for at least two centuries."

—Donald Keyhoe,
Chief of Information of the
American Aeronautics Branch and
retired Marine pilot and major, 1954

Some pointed to Mars's opposition as creating the perfect opportunity for its life forms to scope out the Earth. Sightings tended to spike during those years. One newspaper believed that "the super-intelligent Martians, with their space ships, would choose this special time (once every two years) to hop across, instead of bothering us at all hours of the day and night right through the year." How neighborly of them. "People there could be 50,000 years ahead of us. Space ships would be nothing to them; they could jump the gap of 40,000,000 miles in a mere four minutes, flying with the speed of light," the paper added.

Donald Keyhoe, a retired Marine pilot and major and author of several books on UFOs, agreed with the opposition theory but didn't think they were true Martians. Instead, he suspected that aliens might be using Mars as a base. After speaking with hundreds of pilots who claimed to have witnessed flying saucers of some sort, Keyhoe firmly believed the U.S. government was hiding information from the public. Officials may have feared another Orson Welles-like panic or wished to prevent organized religion from being thrown into upheaval.

In 1958, journalist Mike Wallace interviewed Keyhoe on television. Between commercials for Parliament cigarettes (approved by the United States Testing Company!) Wallace challenged the Major's unique viewpoints and, quoting a columnist, suggested that the sightings might be nothing more than the work of "pranksters, halfwits, cranks, publicity hounds, fanatics in general, and screwballs."

Keyhoe responded that he wished he could show his list of nearly eight hundred witnesses in the aviation industry. "They're still flying and they're still carrying passengers—they've never been grounded," he told Wallace. "They're still guiding airliners in, the radar men are, night after night in bad weather. If they're such screwballs and incompetents why are they still on the job?"

It's possible, of course, that many of those witnesses kept their stories to themselves. At the time, pilots often feared being stigmatized for suggesting anything about aliens. Keyhoe admitted that he hadn't personally seen a UFO. "I've just been a reporter," he said, "and a careful one."

The Air Force didn't care how careful Keyhoe's reporting was. It did its own investigations of UFOs over the decades through a program called Project Blue Book. But by 1969, after 12,618 reported sightings, and 701 remaining in the unidentified category, the book was closed.

The Secretary of the Air Force announced its termination based on its investigations and numerous third-party studies. He offered three conclusions:

1. No UFO reported, investigated, and evaluated by the Air Force has ever given any indication of threat to our national security.

2. There has been no evidence submitted to or discovered by the Air Force that sightings categorized as "unidentified" represent technological developments or principles beyond the range of present-day scientific knowledge.

3. There has been no evidence indicating that sightings categorized as "unidentified" are extraterrestrial vehicles.

CHAPTER 33: UNIDENTIFIED FLYING OBJECTS

Oddly enough, just a year before shutting down Project Blue Book the Air Force included a chapter on UFOs in its *Introductory Space Science* textbook that acknowledged that unexplained sightings offer "the unpleasant possibility of alien visitors to our planet, or at least alien controlled UFOs." Such an admission leads to an obvious question: Why haven't they said hello? The book offered three answers:

1. We may be the object of intensive sociological and psychological study. In such studies you usually avoid disturbing the test subjects' environment.

2. You do not contact a colony of ants—and humans may seem that way to any aliens. Variation: a zoo is fun to visit, but you don't "contact" the lizards.

3. Such contact may have already taken place secretly, and may have taken place on a different plane of awareness—and we are not yet sensitive to communications on such a plane.

The 1970 edition of the textbook was altered to reflect the conclusions of Project Blue Book.

Perhaps the closure of Project Blue Book put some people at ease. But for many, the mystery only deepened. Conspiracy theories about Washington cover-ups continued to swirl, particularly in regard to Roswell. So in 1997 the Air Force made one more attempt to put the fifty-year-old case to rest. Colonel John Haynes, deputy chief of the Air Force Declassification Review Team, held a press conference at the Pentagon and announced the bodies that people had witnessed were nothing more than crash test dummies being used for parachute tests. This was the major finding in a 224-page study called "The Roswell Report, Case Closed." The explanation had one curious flaw: crash test dummies weren't in use until 1953. So how could they have been found in 1947? When reporters challenged him on it, Haynes said that witnesses probably got their years confused. Roswell happened in 1953, not 1947, he maintained.

"Someone said they think the story of the dummies was put out by dummies," a staff member at the International UFO Museum and Research Center in Roswell told newspapers after the press conference. "I think it's another cover-up."

Journalist and producer Lee Speigel, who has reported on UFO sightings for decades and arranged the first United Nations conference on the subject in 1978, found Haynes's explanations absurd, particularly since so many newspaper headlines from that July 8, 1947, date couldn't have been confused about the year. "It's not case closed," he says. Instead, Speigel suggests the federal government has been pursuing a different strategy. "The Air Force is waiting out everyone to be dead. Until there are no living witnesses of the Roswell crash."

HEY MARTIANS, WE'VE GOT
BLINKING LIGHTS, TOO!

The turn-of-the-century scientists had their work cut out for them when they devised their elaborate plans to signal Mars. Ray Stanley was no mathematician or professor, but he had a distinct advantage when he set up his own signals to greet Martians in 1973: They weren't fifty million miles away. Instead, they were right here on Earth, flying by in their fancy saucers with colorful blinking lights.

Stanley and fifteen UFO-believing volunteers arranged a hundred outdoor flood lights in a circle—or saucer shape, if you will—on a rocky hillside near Lake Travis, north of Austin, Texas. They rigged the lights to blink just like the aliens, in every color of the rainbow, visible to 150 miles in the sky. What alien wouldn't be curious? Stanley hoped to at least get a photo of a flying saucer investigating from above or, better yet, a landing and an interplanetary meet-and-greet.

The ambitious project cost him a mere $5,000. "If the government had gone to the trouble to set up this kind of thing, it would have cost about $1 million," Stanley told the press. This way he was able to save the government money and spare them the trouble of orchestrating any elaborate cover-ups.

Stanley and his team had formed a group called Project Starlight International and plotted UFO sightings on a chart over the previous five years. "One curve indicates a rash of sightings about every twenty-six months, which people immediately notice as the Martian approach cycle," he explained. "Maybe ten percent of the people would say there is no such thing as a UFO, but many people want to believe and many people want to see."

A MODERN-DAY STRANGER
IN A STRANGE LAND

Going to Mars or witnessing a UFO isn't the only way to encounter a Martian. Sometimes you can give birth to one. Just like the parents of Boriska Kipriyanovich did.

The young Russian-Martian, born in 1996, claims to have once been a pilot on the Red Planet. Boriska, a child prodigy who began reading, drawing, and painting by the age of eighteen months, started sharing stories of his past life as a Martian at age seven. Scientists were impressed with his knowledge of the universe, planetary systems, and the details of his alleged Martian life. His parents claim no one taught him any of it. (Perhaps he was just a voracious reader?)

Thousands of years ago during the Martian heyday, Boriska says their advanced society enjoyed such luxuries as teleportation and time travel. With triangle-shaped planes they could even fly to Earth. Boris personally flew to Egypt, where he says many secrets are still waiting to be unlocked at the Great Pyramid of Giza and the within the Sphinx.

"The human life will change when the Sphinx is opened, it has an opening mechanism somewhere behind the ear; I do not remember exactly," he said. If only he could jog his memory by using an old Martian time machine and returning to ancient Egypt.

As for the Martians themselves, Boriska says they generally reach heights of seven feet, much like Oomaruru told Dr. Hugh Mansfield Robinson in the 1920s. They are also immortal and cease maturing past the age of thirty-five. A nuclear war, however, negated the powers of immortality for most Martians, although some survived and still live underground, where they're able to breathe carbon dioxide and communicate with other galaxies and planets. He, along with others, was sent to Earth to help save humanity from a similar catastrophe.

So far so good—whether you believe him or not.

n the meantime, the U.S. government has continued its research into the UFO phenomenon. The operation had remained covert until the *New York Times* made it public in 2017, reporting that the Pentagon had quietly budgeted $22 million toward the study of flying saucers—more specifically known as the Advanced Aerospace Threat Identification Program. Harry Reid, the former senator from Nevada, requested funding for the program in 2007 when he served as Senate majority leader. The small, secretive group studied videos of encounters between military aircraft and unknown objects. As Reid told the *Times*:

> *"I'm not embarrassed or ashamed or sorry I got this thing going I think it's one of the good things I did in my congressional service. I've done something that no one has done before."*

The Department of Defense had never acknowledged the program's existence before the 2017 article. It claimed that funding ended in 2012, though supporters within the Pentagon say the search continues. For now, the only aliens that we can be sure of are the robots we've sent to Mars—and if anyone is going to attack our planet, it'll be those of us living on it.

RADIO 'WAR' P
U.S. TO SCAN

FORM PLAN
EXCHANGE

Victory for Lehman Is Declared Certain; 3 on Ticket in Doubt

BAN
FIRE

CHAPTER 4
MARS INVADES
POP CULTURE

York Post

DAY, OCTOBER 31, 1938

MEMBER OF THE ASSOCIATE

NIC BRINGS IN
BROADCAST S

Partial Script of Radio Drama
That Spread War Fear in U.

LAY
AKE
OLL

ctim
in

For thousands of years the mysterious Red Planet has inspired wonder and imagination across the globe, creating hope and anticipation of greater intelligence and answers to age-old questions. Then, at eight o'clock on October 30, 1938, the Martians invaded and ruined everything.

Like most evenings, on this fateful night, families were gathered around the radio twiddling the dial to find a delightful selection of music, comedies, and dramas. But when the dial reached the Columbia Broadcasting System, here's what listeners heard:

> *"Ladies and gentlemen, I have a grave announcement to make. Incredible as it may seem, both the observations of science and the evidence of our eyes, lead to the inescapable assumption that those strange beings who landed in the Jersey farmlands tonight are the vanguard of an invading army from the planet Mars. The battle which took place tonight at Grover's Mill has ended in one of the most startling defeats suffered by an army in modern times; seven thousand men armed with rifles and machine guns pitted against a single fighting machine of the invaders from Mars. One hundred and twenty known survivors. The rest strewn over the battle area from Grover's Mill to Plainsboro crushed and trampled to death under the metal feet of the monster, or burned to cinders by its heat ray."*

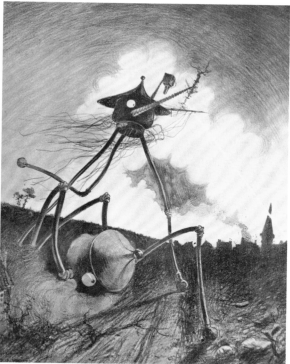

Right: A Martian fighting machine as illustrated by Henrique Alvim Corréa in a 1906 edition of *The War of the Worlds*.

Those who tuned in at the beginning of the program knew they were listening to the Mercury Theatre on the Air's adaptation of H. G. Wells's *The War of the Worlds*, directed by and starring twenty-three-year-old Orson Welles (*above*). Those who were busy listening to the end of ventriloquist Edgar Bergen and his dummy, Charlie McCarthy, tuned in late and missed that all-important introduction. Many believed the invasion was real, triggering a nationwide panic.

All those theories from the world's top astronomers and scientists that newspapers had published over the decades finally seemed to be happening. The public had been waiting for this moment, but they expected the friendly, canal-digging brand of Martians that Percival Lowell and Camille Flammarion spoke

of, or the scientific geniuses that Nikola Tesla thought were trying to contact us. What they did not expect was belligerent octopus-like evil creatures armed with deadly heat rays. But that's how Wells had described them back in 1897, when his serialized story was first published. And that is who was arriving.

Wells had been paying attention to all the Martian hullaballoo toward the end of the nineteenth century and even name-dropped Giovanni Schiaparelli early in the story. Given the information he had at the time, life on Mars seemed perfectly possible. In fact, he wrote an essay entitled *Intelligence on Mars* in 1896, in which he described Mars as being very similar to Earth, with its land and oceans, continents and islands, mountains and seas, and pleasant weather. But the similarities ended there.

"Granted that there has been an evolution of protoplasm upon Mars," Wells wrote, "there is every reason to think that the creatures on Mars would be different from the creatures of earth, in form and function, in structure and in habit, different beyond the most bizarre imaginings of nightmare."

The Martian attackers of *The War of the Worlds* parallel an all-too-common story on Earth: a powerful civilization tries to conquer a weaker one. Only in this case, the intruders ultimately are defeated by germs. Wells's tale set the standard for the Martian invasion genre and inspired the work of future sci-fi writers. But despite the book's popularity, plenty of people never read it—as Orson Welles clearly proved forty years later.

The radio adaptation shifted the action from England to the United States and began the story with several news alerts interrupting the "music of Ramón Raquello and his orchestra." These types of bulletins were something listeners had gotten used to, particularly in recent months as World War II was brewing in Europe. So, nothing seemed strange or out of the ordinary when reports came in from Professor Richard Pierson of the Observatory at Princeton, Professor Morse of McGill University in Montreal, Professor Farrell of the Mount Jennings Observatory in Chicago, and other people who sounded very official and awfully important.

Initially, the audience simply heard reports of explosions witnessed on the surface of Mars and things didn't seem so bad. Professor Pierson described "transverse stripes" seen through his telescope but assured the audience they were not the legendary canals and that the chances of intelligent life on Mars were a thousand to one. Another report described a meteor strike in New Jersey, and suspense began to build. As the story unfolded, the skittish public learned that it was no meteor, but a metallic spacecraft packed with destructive Martians. By then, the actors portraying panicked citizens were being outdone by their listeners.

Reporters surrounded Orson Welles the day after his *War of the Worlds* broadcast, 1938.

People grabbed whatever personal belongings they could and swarmed the streets, being careful to cover their faces with towels to protect themselves from gas that might be spewing from the meteor. Delirious motorists shouted at each other to "drive like hell into the country" and warned that "we're being bombed by enemies." In New York City, hundreds of doctors and nurses mobilized to help potential victims.

Churches around the country were flooded with people dropping to their knees and praying for help. An Indianapolis woman rushed in screaming, "New York destroyed; it's the end of the world. You might as well go home to die. I just heard it on the radio." The services ended on the spot for anyone wishing to take her advice.

A sixty-year-old Baltimore jeweler ran to his window to look for signs of the invasion. The fear was so real that he suffered a heart attack. Two weeks later it proved fatal.

San Francisco reported hundreds of calls to police stations from men ready to sign up for the fight. "My God! Where can I volunteer my services?" one of the panicked men asked.

"We've got to stop this awful thing!" Police, too, were confused.

Not only did all these people miss the announcement at the start of the broadcast, but they also missed the four reminders spread throughout the program: "You are listening to a CBS presentation of Orson Welles and the Mercury Theatre on the Air in an original dramatization of *The War of the Worlds* by H. G. Wells." They were too busy running for their lives until the message finally spread.

One New Yorker described the revelation: "When I got to the street there were hundreds of people milling around in panic. Most of us ran toward Broadway and it was not until we stopped taxi drivers, who had heard the entire broadcast on their radios that we knew what it was all about."

Switchboard operators at newspapers helped, too. They got so sick of all the calls that they stopped bothering to say, "Hello," and immediately calmed callers with the news that it was just a radio show.

Welles ended the program clarifying the situation once again:

"This is Orson Welles, ladies and gentlemen, out of character to assure you that The War of the Worlds *has no further significance than as the holiday offering it was intended to be. The Mercury Theatre's own radio version of dressing up in a sheet and jumping out of a bush and saying, Boo! Starting now, we couldn't soap all your windows and steal all your garden gates by tomorrow night. . . so we did the best next thing. We annihilated the world before your very ears, and utterly destroyed the CBS. You will be relieved, I hope, to learn that we didn't mean it, and that both institutions are still open for business. So goodbye everybody, and remember the terrible lesson you learned tonight. That grinning, glowing, globular invader of your living room is an inhabitant of the pumpkin patch, and if your doorbell rings and nobody's there, that was no Martian. . . it's Hallowe'en."*

Throngs of people awaited Welles as he left the recording studio and made his way to the lobby. "Someone shoved a telephone into my hand and some fellow from out in New Jersey was telling me that emergency squads were organized and that the women were all herded into the churches and what in God's name would he do next?" Welles described in an interview several days after the ordeal. "Then came another call, a woman's voice shouted: 'It's hell! Oh, it's hell!' You can well imagine that just about that time I was agreeing with her."

Angry listeners wrote threatening letters to Welles, calling him "depraved" and "despicable" and challenging him to fights. "Holy smoke, I've even got people coming to New York to kill me," Welles said. He tried to make things right when he could, like offering to pay a man $2.50, the amount he'd blown on a train ticket to the country instead of buying the new shirt he was saving up for.

Many listeners who had paid closer attention and sat peacefully in front of their radios were shocked that so many people were duped. In a letter to the editor of the *Washington Post*, one woman expressed her disgust over Americans' stupidity. "Does no one in the United States read?" she asked incredulously. "I can't imagine an intelligent person who has not read Wells's story and thrilled to the excitement in that highly imaginative piece of fiction. Even people who never buy a book because they already have one should be able to read the 'Radio Times Table'; and if they can't read and are listening to a radio program, can't they hear when an announcer says this is a 'play'? As an example of gross ignorance and pure, childish stupidity, Sunday night's exhibition beats everything in the paper for weeks. The people who were 'angry' were the most ridiculous of all— one is always angry when one is made to look foolish."

As for H. G. Wells, the author didn't panic about Martians, but he was not

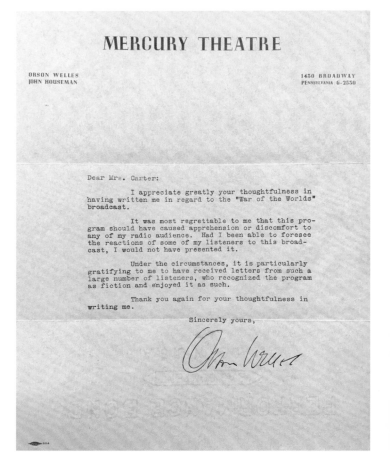

pleased with the panic his story caused. He called Welles's version an "outrage" and demanded his American agent investigate the broadcast. That said, the radio show boosted sales of his book.

The Federal Communications Commission looked into the case as well but, after five weeks decided not to take action against CBS. However, it did regret that some listeners "mistook fantasy for fact" and assured those listeners that CBS would never again try to fool them with creative news broadcasts.

The Martian panic of 1938 has become the stuff of legend, but was it truly as bad as reported? A 2008 Radiolab podcast suggested that newspapers may have overhyped the event to make radio look bad. The relatively new medium had cut into their advertising revenue and was hurting their profits. So, how about showing those advertisers that radio couldn't be trusted? If radio could be damaging to the public, it'd be best to keep those ad dollars with newspapers.

A scathing editorial in the *New York Times* on November 1, 1938, supports the theory:

> *"Radio is new but it has adult responsibilities. It has not mastered itself or the material it uses. It does many things which the newspapers learned long ago not to do, such as mixing its news and its advertising. Newspapers know the two must be rigidly separated and plainly marked. In the broadcast of The War of the Worlds blood-curdling fiction was offered in exactly the manner that real news would have been given and interwoven with convincing actualities, such as an ordinary dance program, a definite locale and the titles of real officials. Radio officials should have thought twice before mingling this news technique with fiction so terrifying. Horror for the sake of the thrill has been legitimately exploited on the air."*

It would have been quite a coordinated attack by the nation's newspapers to conspire against radio so quickly. Regardless, the broadcast created a stir, and given the mass or not-so-mass fright, it would seem like the kind of thing that would never happen again. Until it did—eleven years later.

FLASH, AH AH,
HE'LL SAVE EVERY ONE OF US . . .
FROM THE MARTIANS

Where there's a panic, there's a buck to be made. On November 7, 1938, one week after *The War of the Worlds* radio broadcast, Universal rushed an abbreviated version of *Flash Gordon's Trip to Mars* into theaters.

The fifteen-chapter serial had been released eight months earlier. Now called *Mars Attacks the World*, it featured a typical, but timely, Flash Gordon storyline: a journey into space to save Earth. Flash races to the Red Planet and, with the help of Martian clay people, battles Ming the Merciless and his ally, Queen Azura of Mars, to prevent their death ray from destroying the world.

Less scary for America, and more money for Universal.

THE WAR OF THE WORLDS PART 2: ATTACK IN ECUADOR

On February 12, 1949, two radio producers in Quito, Ecuador, were looking to make a splash on air. A local panic apparently seemed like the perfect plan, so they broadcast their own version of *The War of the Worlds*. This time, the fictional Martians ended up killing fifteen real people and injuring many more.

Unlike the CBS version, the Quito producers made no announcements at the beginning to let listeners in on the dramatization. Instead, they interrupted one of Ecuador's most popular bands with an important message: "Here is an urgent piece of news. Listen—and tell your neighbors. I have a grave statement to read to you, fellow citizens. Our country has been invaded for an hour by men from Mars, who approached the Earth in a space ship, and landed near Cotocollao."

Listeners were informed that the tentacled Martians had already used their death rays to destroy a neighboring town twenty miles to the south and were heading straight for Quito. The army had been battling the creatures and failing miserably.

An actor impersonating the Minister for Defense ordered martial law and pleaded with citizens to remain calm (a tall order for people already panicking). The situation was exacerbated by the next actor, who portrayed Quito's mayor. "Let the women and children flee to the surrounding heights, so that the men are left free to fight to the last bullet, to the last breath," he insisted. Hearing a "priest" begging God for mercy didn't help either.

The horrifying reports continued and the terrified listeners fled their homes, shouting, crying, and praying in the streets. An Englishman visiting Quito at the time described the scene: "Men and women of all ages, mothers with children on their backs and in their arms, all frantic with desperation, were choking the narrow passages, trying to reach the safe road to San Domingo. Costly chauffeur driven limousines were held up by flocks of sheep that the owners wanted to take with them. Priests in purple robes assisted the old and the infirm."

Even the police were fooled. Units had immediately been sent to Cotocollao to assist the supposedly besieged army.

Once the producers realized what was happening outside the studio, they made an announcement to clear the air: "This is only a broadcast—not reality. Nothing has happened. Go home." It was too late. Everyone was busy running for their lives.

Like the panic caused by Orson Welles's broadcast, the message eventually spread through the streets. But in Ecuador, people's fear turned to rage, and their rage turned into a mob. They headed straight to Radio Quito, where they threw rocks and set fire to the building. Nearly a hundred people were trapped inside; many on the third floor were forced to jump. Firefighters struggled to reach the station due to the masses of people crowding the streets.

In the aftermath of the riots, both radio producers lost their jobs and were indicted. One reportedly fled the country. The other was eventually exonerated. According to his daughter, the event was a bittersweet experience. "He was proud of it because it was such a good artistic production that people believed it — and that was his job as a radio actor," she said. "But it was a tragedy because people died."

EARLY INVASIONS AND BARSOOMIAN BATTLES

The War of the Worlds is perhaps the most famous tale of Martians, but just as life on the Red Planet had intrigued the minds of many nineteenth- and twentieth-century scientists, so, too, had it captured the imagination of writers. Astronomers couldn't see life, but authors had little trouble creating it on paper.

Percy Greg, for example, populated Mars with a race of fair-skinned bearded dwarfs, unicorn-like creatures, gigantic squirrels, dragons, and other strange beasts in his 1880 book, *Across the Zodiac: The Story of a Wrecked Record.* In it, the narrator builds a spaceship and flies to Mars, where he meets the little Martials (yeah, people used to say "Martial") and finds himself wielding a sword in the middle of political and social strife.

Greg gave the Martials their own language—considered the first alien diction in fiction—and described their stylography and grammar rules in detail. More notably, he invented a new word in English: Astronaut. Used as the name of the ship, the term had not been written or spoken before. *Across the Zodiac* has been largely lost to obscurity, but it lives on with a forty-two-mile cavity on Mars, which NASA named the Greg Crater in 2010.

Unlike Greg's dwarfish and vocal people, Robert D. Braine envisioned Martians as standing ten feet tall and communicating through color-coded eye contact instead of speech. In his 1892 book *Messages from Mars by the Aid of the Telescope Plant*, Braine's creatures lack mouths but have white horns on their foreheads and three-foot-long noses that suck nutrients from an elixir-filled reservoir. Telescopes connect them with Earth through a magical sacred plant found on an uncharted island in the Indian Ocean. The messages received from this telescopic plant inform the islanders that the Martians call their planet Oron, and that having eliminated disease, war, money, private property, cities, canals, food, stores, fashions, accessories, barbers, and more, they have found paradise. Though that seems to be a matter of opinion.

> "ALL THAT COLUMBUS CAN HAVE FELT WHEN HE FIRST SET FOOT ON A NEW HEMISPHERE I FELT IN TENFOLD FORCE AS I ASSURED MYSELF THAT NOT, AS OFTEN BEFORE, IN DREAMS, BUT IN VERY TRUTH AND FACT, I HAD TRAVERSED FORTY MILLION MILES OF SPACE, AND LANDED IN A NEW WORLD."

—Narrator
from Percy Greg's
Across the Zodiac, 1880

Gustavus W. Pope agreed with scientists who claimed Martians were superior in intelligence. In *Journey to Mars* (1894), his human-like life-forms were so smart they created "ethervolt cars," anti-gravity aircraft, and communications gadgets similar to television and video chat.

The unknown of Mars made it rich territory for the most imaginative stories. Writers could borrow as much science as they wanted and project whatever kind of world they desired. In 1893's *Unveiled in Parallel: A Romance*, the Red Planet became a utopia for its authors, listed only as "Two Women of the West." At a time when women had few rights, their Mars was a place where all women were equal, if not superior to men, with all the freedoms they deserved. The authors were later identified as Alice Ilgenfritz Jones and Ella Merchant of Cedar Rapids, Iowa.

When the male protagonist arrives on Mars, he is immediately impressed by every wondrous detail—from the cold, sparkling drinking water to the mosaic floors and marble statuettes that adorn Martian mansions. The Martian people are even more impressive. Severnius and Elodia, a brother and sister, welcome the traveler into their home, where conversations quickly illuminate the many differences between the planetary species. The power Elodia holds in society is the most striking.

"I do not know how many great enterprises she is connected with," Severnius explains to him, "railroads, lines of steamers, mining and manufacturing. And besides, she is public-spirited. She is much interested in the cause of education—practical education for the poor especially. She is president of the school board here in the city, and she is also a member of the city council. A great many of our modern improvements are due to her efforts." She sounds more like a modern woman than a Martian.

Jones and Merchant, our two women from the West, state their plight through the Earthman's response: "We humor them, patronize them, tyrannize over them. And they defer to, and exalt us, and usually acknowledge our superiority."

As their conversation continues, Severnius reveals that Martian women earn three times as much as their human counterparts and that they have equal rights, including suffrage—a privilege that was still twenty-seven years away from becoming an amendment to the U.S. Constitution.

"We do not hold that women are our political equals," the Earthling tells Severnius, adding that if they try to vote, "we simply throw their ballots out of the count."

"That seems to me a great unfairness," Severnius responds, speaking for Jones, Merchant, and the entire female population of nineteenth-century America.

In 1897, 114 years before Andy Weir wrote *The Martian*, George Du Maurier wrote his own book with that title. The story follows two friends, Maurice and Barty, from childhood to adulthood. Barty, the reader learns, has been inhabited by a Martian called Martia, who helps him become a famous writer. Martia, Du Maurier writes, "had gone through countless incarnations, from the lowest form to the highest, in the cold and dreary planet we call Mars. . . . Man in Mars is, it appears, a very different being from what he is here. He is amphibious, and descends from no monkey, but from a small animal that seems to be something between our seal and our sea-lion."

THE
MARTIAN

DV
MAVRIER

MARS
SELLS

Scientists had their theories on the matter, but no one knew what Martians might look like. However, there was one thing everyone could agree on: they'd better smell good. In 1893, Kirk's American Family Soap leveraged hype about Martians and Charles T. Yerkes's new refracting telescope, the largest in the world, to hawk its products. Eight years later, Pears' Soap capitalized on Nikola Tesla's promise that contact would happen any day.

The Arlington Chemical Company in Yonkers, New York, took things even further to promote its Liquid Peptonoids. The druggists produced an eighteen-page pamphlet called the *Mars Gazette* and claimed that it was an exact replica of an original straight from the Red Planet. Dr. C.B. Hustler, the company explained, had brought it back from his trip and was kind enough to translate it into English. The Peptonoids peddler had already sold his goods to all of Earth's doctors and found a new needy audience on Mars. Even King Flammarion, the "Supreme and Mighty Ruler of the Red Planet" who was "tortured day and night by the Incubus of Disordered Nutrition," was soothed by Dr. Hustler's remedy. The *Gazette* proclaimed Liquid Peptonoids to be "unequalled for recruiting the energy of invalids and convalescents." If nothing else, they were surely successful in flattering French astronomer Camille Flammarion.

All these Martians had redeeming qualities on different levels: intelligence, inventiveness, productivity, and, to varying degrees, humanness. Then H. G. Wells came along and turned them into a bunch of grumpy Earth-destroying monsters. But Garrett P. Serviss was not going to stand for that.

The American astronomer and writer took it upon himself to seek a swift revenge on Mars for its attack in *The War of the Worlds*. To do so, Serviss created a fictional super team composed of the greatest minds at the time, including Thomas Edison, physicist Lord Kelvin, and X-ray inventor Wilhelm Roentgen, and wrote an unofficial sequel to Wells's novel. He called it *Edison's Conquest of Mars*. Today, we'd probably send Elon Musk, Jeff Bezos, and Richard Branson on this vengeful mission. (Actually, they could send themselves, as we'll see in the next chapter.)

The serialized story appeared in William Randolph Hearst's newspapers in 1898, striking while the iron was hot. Serviss explained that some of the invading Martians had survived the germs and fled back to Mars. Edison, having studied their spacecraft, "invented and perfected a flying machine much more complete and manageable than those of the Martians had been." It could zip around, turn, rise, and fall with "the quickness and ease of a fish in the water." His fellow scientists developed equally impressive weapons to fight the Martians. The story is particularly noteworthy for containing the first battle in space, the first alien abduction in fiction, and for the proposal that the Martians visited Earth thousands of years ago and built the pyramids and sphinx in Egypt. Though Thomas Edison is the hero of the novel, the famous inventor did not participate in its creation—nor did H. G. Wells.

Mr. Edison's Immediate Purpose Was Fulfilled and We Hastened Back to the Earth.

A Little Gaseous Globe Darted Into the Upturned Face of the Martian.

EDISON'S SILENT TRIP TO MARS

Thomas Edison didn't collaborate on *Edison's Conquest of Mars*, but he may have been flattered by his heroics portrayed in the story. If not, he was at least inspired to create his own piece of Mars fiction. His five-minute silent film, *A Trip to Mars*, released in 1910, features a scientist who develops a method to reverse gravity and floats his way to the planet next door. There, he encounters giant Martians and becomes their plaything before escaping and jumping back to Earth.

Today it's recognized as the first American-made science-fiction movie.

Edison and his fellow geniuses prevailed (the title is sort of a spoiler, isn't it?), but they might've enjoyed a little help from a fictional contemporary, John Carter. Edgar Rice Burroughs dreamed up the adventurous character in 1911 after his own series of earthly exploits all across America as a member of the U.S. cavalry and stints as a cowboy, a storekeeper, a gold miner, a cop, an untrained accountant, and a pitchman for pencil sharpen-

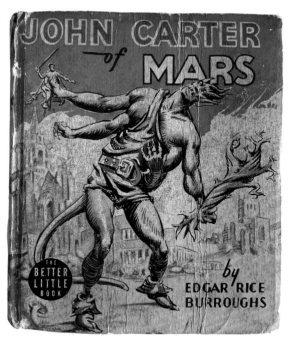

ers. During downtime at the pencil sharpener gig, he picked up a sharp pencil and started to write.

When the first chapter of *Under the Moons of Mars* ran in the February 1912 edition of *All-Story Magazine*, Burroughs's name was listed as Norman Bean. He had submitted his manuscript under the name Normal Bean to clarify that he was just a regular guy—not some wacko living in a fantasy world. The editor assumed "Normal" was a mistake and changed it to "Norman." The story's swift success convinced Burroughs to drop the Bean and take the credit.

After all his struggles with finding steady work to support his family, the former miner struck gold on Mars. Finding work would never again be a problem.

Burroughs drew inspiration from Percival Lowell and opened the first chapter of the Barsoom saga ("Barsoom" is Martian for "Mars") with Carter, a Civil War vet, hiding out in an Arizona cave. Surely it wasn't far from the astronomer's observatory hideout.

Carter is transported from the desert to Mars by simply gazing into the night sky, closing his eyes, and longing for the distant red dot. When he awakens, he's magically there. Burroughs brought Lowell's idea of a dying planet to life with brutish green Martians, eight-legged thoats, giant white apes, ferocious Warhoons, and, of course, plenty of canals. Carter is quickly thrust into action as he tries to save a beautiful humanoid princess from the clutches of the Tharks—giant in stature and equipped with four arms wielding a horde of weaponry. Luckily for our Earthman, he quickly discovers that he has a distinct advantage: superpowers in their low-gravity world. A simple leap launches him more than fifty feet in a single bound, and his fists of fury can knock out the mightiest of Martians.

Originally written in serial form, the story kept readers on the edge of their seats with swashbuckling battles, rousing captures, daring escapes and rescues, interstellar romance, and cliffhanger after cliffhanger. Burroughs's Barsoomian tales continued through a series of eleven novels that were heavy on fantasy and action and light on science. Over the next hundred years, his vision of Mars would entertain millions, including several future authors who were inspired to spread adventure across space through their own beloved characters and films, including *Buck Rogers*, *Flash Gordon*, *Star Wars*, and more recently, *Avatar*. Science fiction owes much to the John Carter saga, even though at the time the term "science fiction" hadn't yet been invented. Hugo Gernsback, another writer with a powerful imagination, soon took care of that.

MARS
SATIRIZES EARTH

Someday, in the not-too-distant future, honeymooners may skip Maui in favor of Mars—just like they did in W. Clyde Spencer's 1911 comic strip, "A Trip to Mars." The short-lived series depicted newlyweds living it up on the Red Planet alongside friendly, big-headed, tentacled Martians. Spencer's political, social, and cultural humor all set up the same punchline about how different life is ninety-five million miles away from home.

SCIENCE FICTION AND THE EVOLUTION OF IDEAS

Hugo Gernsback was not a scientist by trade, but as a writer and publisher of numerous science magazines, such as *Science and Invention* and *Amazing Stories,* he influenced the likes of Jack Parsons and so many other readers with his unique vision of what was to come. Fiction served as a vehicle for scientific predictions, which he was pretty good at. Gernsback foresaw television, video chat, radar, air conditioning, tape recorders, sky writing, fluorescent lights, and other things we're glad he thought of. He coined this type of writing "science fiction."

In his youth, Gernsback discovered Percival Lowell's books on Mars and fell in love with science and the possibilities it offered. Fascinated by the prospect of life on the Red Planet, he sought to contact existing intelligent life as early as 1909. Like Tesla, he hoped to send a super-signal from a wireless transmitter—one

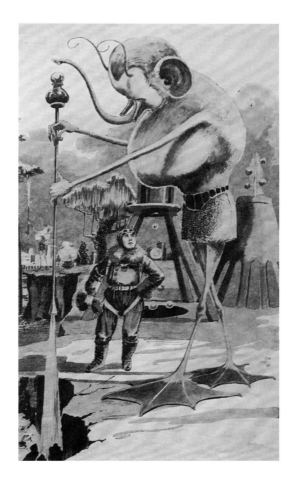

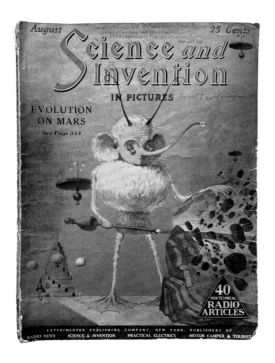

that required every wireless station in America to hook into one key location in Lincoln, Nebraska. By the start of the 1920s, he wanted to dispatch blinking messages through space using a thousand powerful searchlights.

Gernsback never made contact with Martians, so he decided he'd just make his own. He formulated his concept of what a Martian might look like based on his knowledge of evolution here on Earth and wrote about it in several of his magazines. It was a vision that resembled the one Edmond Perrier described just a few years earlier: unusually tall, large ears, inflated chest, skinny legs, and an enormous head to hold a large, superior brain.

Over the millennia, the atmosphere had slowly vanished, leaving little oxygen for the modern Martian. Gernsback believed these beings would've developed an enlarged chest to survive. He backed this claim by pointing to the Cholos Indians in the Peruvian mountains who live at an altitude of 12,000 feet. Their chests measured five inches more than the average of people living closer to sea level.

Mars's gravitational pull, being lighter than Earth's, meant that Martians could jump farther and lift greater weights than any human being. Gravity would lead to creatures standing fifteen to twenty feet tall, and the thinning atmosphere would have conducted sounds and scents poorly, causing them to grow "large bat-like ears" to hear better and long noses to reach out to odors. "Just as the elephant had to grow a long trunk in order to make it easier for him to get to his water and just as the giraffe has a long neck to reach the food he likes," Gernsback wrote, adding:

> "Having attained a far more advanced civilization, performing all work by machines and hardly ever attempting manual labor, the Martian's arms have shrunk until they are little more than bones with skin covering them. The body weighing much less on Mars, and the Martian probably moving around only in mechanical contrivances, his legs have become almost useless and are therefore similarly attenuated. They also have but tiny muscles covered with skin. But in order to support such a tall body, the Martian must have large feet.

> "We may expect to find the Martian with projecting eyes if our deductions are correct and the temperature on Mars being nearly always freezing even at the Equator and going below zero at higher latitude, the Martians will probably be covered with thick fur or feathers in order to keep him warm. The two horn-like projections on his forehead are antennae, and constitute the Martian's telepathic organ."

This telepathic ability, as ascribed to Martians by Dr. Hugh Mansfield Robinson, turn-of-the-century Spiritualists, and future science fiction writers, was explained by Gernsback as a necessity that was due to the difficulties of speech—again, because of sound traveling in the rare atmosphere. Ants communicate telepathically, in a sense, so why couldn't Martians develop the same powers?

Had he been around, Lowell would have loved these Martians and appreciated Gernsback's scientific reasoning behind their lifeform. Movie studios, by contrast, loved everything about Martians except perhaps Gernsback's detailed description. Special-effects technology was too primitive, and even if filmmakers had had the capabilities, they couldn't have afforded those types of intricate creatures. That's why they were called B movies.

Mars was a muse for many early writers. Isaac Asimov, often considered one of the "Big Three" science-fictions writers, along with Robert A. Heinlein and Arthur C. Clarke, called Stanley Weinbaum's *A Martian Odyssey* (1934) a genre-changing work. In the 1950s, Donald A. Wollheim found an eager sci-fi audience in teenage boys. And Leigh Brackett later went on to write the first draft of *The Empire Strikes Back*.

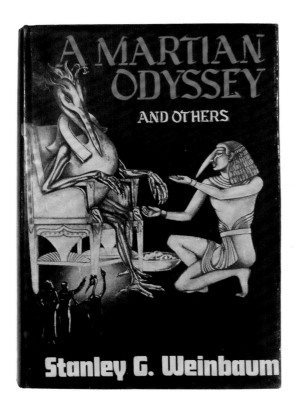

A MARTIAN ODYSSEY AND OTHERS — Stanley G. Weinbaum

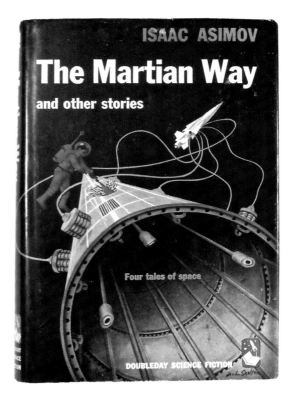

ISAAC ASIMOV — The Martian Way and other stories — Four tales of space — DOUBLEDAY SCIENCE FICTION

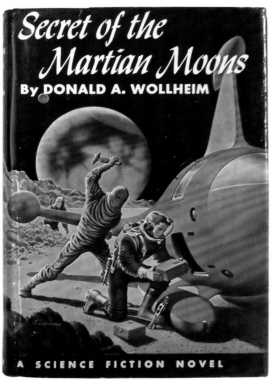

Secret of the Martian Moons — By DONALD A. WOLLHEIM — A SCIENCE FICTION NOVEL

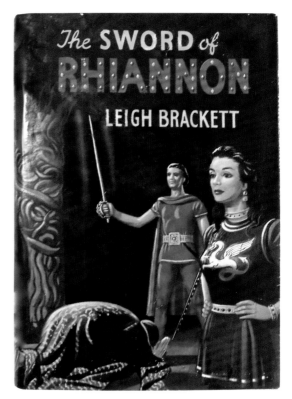

The SWORD of RHIANNON — LEIGH BRACKETT

FOURTH SHLOCK FROM THE SUN

Americans had been promised Martians since the turn of the century and, doggone it, they were going to get them. Radio signals hadn't done squat, and the invasion of 1938 was a false alarm. The UFO sightings of the later 1940s through the 1960s offered hope and fear to those looking for either or both. But the only place Martians had landed was on the big screen. Movie producers with ample imaginations, tiny budgets, and a week or two of shooting delivered all types of Martians and space missions to theater-goers.

Despite their low quality, some of these B movies attempted to gain credibility by building off the scientific ideas of Lowell, Flammarion, and others. *Flight to Mars* (1951), for example, portrayed the Red Planet as a dying world populated by a technologically advanced people living underground. Life may indeed flourish underground, but since microbes don't make for exciting plots, this film's Martians look just like Earthlings modeling slightly different fashions. The human explorers are captured but join forces with Martian sympathizers to escape.

The Wizard of Mars (1965) loosely borrows its story from *The Wizard of Oz*, but without the charm, characters, storyline, drama, wonder, or, well, anything good. What it does offer is four explorers (including a woman named Dorothy) wading through the infamous Martian canals before following a golden road to an ancient city that is frozen in time. Dorothy and her three friends save the Martians by unfreezing time and then, taking a page from Judy Garland's Dorothy, wake up in their spacecraft and discover the whole thing was a dream.

Other films explored what an advanced civilization on the Red Planet might do if it decided to pay a visit to Earth. In *Invaders from Mars* (1953), that meant taking over the minds of men and women to do their evil bidding. Such invasion and mind control tactics tapped into widespread Cold War fears of Communism and a Russian nuclear attack, but these scary notions were much more entertaining when Martians were the enemy. This time, the Martian is a large, brilliant, bulbous head with human features and small tentacled limbs who sits expressionless in a protective bubble, like the lovechild of a giant light bulb and a squid who's trapped in a display case and bored to death. He's described as "mankind developed to its ultimate intelligence"—yet if this is what ultimate intelligence looks like, most of us might prefer to stick with our plain old average levels of intelligence. Despite all those smarts, this mastermind needs the help of large green humanoid brutes to carry out his malicious plan, which begins once his flying saucer lands in the backyard of little David MacLean's house.

David, a boy of about twelve, spots the UFO and tries to get help from his parents. But they, and any other adult who investigates the site, fall into a sand trap and are overpowered by the green goons, allowing the Martian genius to turn them into his human zombies. David finally finds hope in an astronomer and a doctor who believe his crazy tale and recognize that rockets being built nearby may pose a threat to the Red Planet. If that were the case, the scientist reasons, it would make perfect sense for Mars to attack Earth before Earth attacks them.

In the end, the whole fiasco turns out to be nothing more than a bad dream, until Dorothy—er, make that David—looks out of his window again and sees a flying saucer. "Gee whiz!" he says, as the screen fills with a title card announcing "The End." Audiences surely groaned the same thing.

The plot of *Devil Girl from Mars* (1954) stems from a massive gender war on the Red Planet that vanquished any little green men or men of other colors and sizes. This left Mars with a pretty serious reproduction problem, so the Devil Girl, called Nyah, visits Earth in search of fresh meat to bring back with her. Tall and clad in full latex, she looks like a dominatrix who would've had Darth Vader swiping right on Tinder. Unfortunately, Nyah lands near a rural Scottish inn that doesn't give her the best options: a newspaper reporter, a scientist, and an ex-con.

Naturally, the men are curious about Nyah and what on Earth she's doing in the outskirts of Scotland. Turns out she was headed for London, where she would've had a jolly good selection of fellows to choose from, but her entry into the atmosphere sent her plans into a tailspin, and now it'll take four hours for the ship repair itself. So, her search for men begins with them. To her dismay, there are no volunteers to go to a planet filled with lonely, attractive women wearing skin-tight clothing—which may be more unbelievable than aliens landing in the first place.

Mars may have needed men in the 1950s, but things changed drastically on the Red Planet over the next decade, at least according to *Frankenstein Meets the Space Monster* (1965) and the more overtly titled *Mars Needs Women* (1967).

Despite its name, *Frankenstein Meets the Space Monster* does not feature Frankenstein, but that didn't let producers stop it from capitalizing on the popularity of Mary Shelley's monster. As for the Space Monster, it's an angry apish creature accompanying a group of survivors from a devastating atomic war back home on Mars. Led by Princess Marcuzan and Doctor Nadir, who travel in a spacecraft designed like a geodesic dome with landing gear, their mission is to kidnap Earthwomen so they can repopulate Mars. The Princess, being the only Martian female, clearly doesn't want to handle the breeding on her own, but she's perfectly happy to inspect the girls ("turn around") as they're captured from beaches and swinging sixties parties. Doctor Nadir smiles with approvals and appears eager to bring the nefarious operation to fruition. With his bald head and pointy ears, Nadir looks like a Martian predecessor to Dr. Evil from the *Austin Powers* films.

The pair's simple plan runs amok after they shoot down what they believed was a defense missile fired at them. It was, in fact, a rocket headed to Mars piloted by a humanlike android called Frank. Not only does his mission get ruined, but after a battle with a Martian soldier, so does his face. It's half destroyed, turning Frank into a Frankenstein-like monster. He stumbles his way toward the space dome and sets the captive women free before fighting the space monster and killing everyone left aboard.

This means two things: this B movie is mercifully over, and Mars still needs women. Which brings us to the aforementioned and equally crummy *Mars Needs Women*.

MARS NEEDS MOVIE-GOERS

When you've made a bad film, any gimmicks to get people into theaters helps. *Frankenstein Meets the Space Monster* tried to woo audiences with special "Space Shield Eye Protectors" promising to safeguard them against "the high intensity cobalt rays that glow from the screen" and to prevent their abduction to Mars.

In Philadelphia, a film promoter rented a fleet of "out-of-this-world" Mustang convertibles and drove them around town for a week in advance of the film's opening. The cars were filled with "out-of-this-world" blondes and brunettes and the space monster to turn heads.

The film opens with several women disappearing mysteriously—from a tennis court, a restaurant and, gratuitously, a shower. Scientists at the United States Decoding Service soon learn why after translating incoming signals from Martians. The message is just three words: Mars. Needs. Women.

The Martian leader, called Dop, confirms this need through a loud speaker at the USDS: "We have attempted to seize three women by transponder. We have been unsuccessful. Now we come in person. Prepare for materialization."

Once materialized, we see that the Martians look just like ordinary humans with a penchant for black threads, swim caps, and bulky headphones. Dop insists they only journeyed to Earth because of an unfortunate genetic situation on Mars that has resulted in "a preponderance of male births over female" and desperately need some women. But only five. Five unmarried, healthy, fertile women.

Using a divide-and-conquer strategy, one Martian heads straight to the strip club and gawks at a dancer in an unnecessarily extended scene—perhaps to show how great he is at blending in with the men of Earth, or perhaps to fill time in the movie. Using a hypnotic Martian stare, the others cast spells over a homecoming queen, a flight attendant, and an artist. The plan is working perfectly until Dop meets a beautiful scientist who just happened to have written her thesis on space genetics. As the Martian leader discovers his human side, he decides to stay on Earth with his new girlfriend. Mars be damned. Let them repopulate the planet with four women.[10]

---

10 Filled with an abundance of stock footage and poor performances, *Mars Needs Women* later redeemed itself by lending a sample to a remix of 1987's classic "Pump Up the Volume," by British recording act M|A|R|R|S.

MARVIN THE MARTIAN, BELOVED DESTROYER OF EARTH

Since his introduction in 1948, Marvin the Martian has been one of Looney Tunes' most beloved characters—despite the fact that his sole mission is to blow up the Earth. Of course, he has a good reason: it's blocking his view of Venus.

Marvin was a product of the times. The UFO craze had swept across the country, science fiction and comic books were inspired by the space race, and the Cold War created a cloud of fear over enemy invasions and mass destruction. When animator Chuck Jones mixed it all up on paper, it resulted in an ideal new enemy for Bugs Bunny—one he couldn't outwit so easily.

Marvin the Martian debuted in *Haredevil Hare*, where he was introduced as "Commander, Flying Saucer X-2," accompanied by his Martian dog, K-9. Bugs finds Marvin on the moon, preparing to attack Earth with his "Illudium Q-36 Explosive Space Modulator," which sounds impressive but looks like a stick of dynamite.

Unlike Bugs' other main nemesis, Yosemite Sam, Marvin is quite eloquent, despite having no mouth. Aside from his sneakers, the rest of his look is based on the uniform of Mars—the Roman god of war, who wore a crested helmet and armor. Marvin doesn't get quite the same regal treatment though. Instead of a fancy plume atop his helmet, he's got a broom. Perhaps he planned to use it to clean up after destroying Earth.

In 2004, the launch team at Cape Canaveral chose Marvin as its mascot for the *Spirit* Mars Rover. The mission's launch patch design features him standing in front of a patriotic background proudly saluting, suggesting that maybe after all these years he's finally grown to love Earth.

Bad acting and bad-if-not-worse special effects typically turn sci-fi B-movies into comedies, but sometimes they're intended to draw laughs. A little humor is often the best escape from the real world, especially when that threat was a nuclear one. Abbott and Costello and The Three Stooges were happy to oblige.

In 1953's *Abbott and Costello Go to Mars*, the oafish comedians are working aboard a spacecraft and inadvertently fire it up, launching them into the skies. Twists and turns weave them through New York City skyscrapers, in and out of the Lincoln Tunnel, and around the country before the accidental astronauts safely land on what they believe is Mars. Unbeknownst to them, they've arrived on the outskirts of New Orleans during Mardi Gras. As they wander through the bayou into town, they assume costumed revelers wearing large masks are Martians. Typical Abbott and Costello shtick ensues from there.

By early 1962, the Cuban Missile Crisis was brewing and fears of a nuclear attack grew worse. Moviegoers traded the Red Scare for the Red Planet and an onslaught of absurdity in *The Three Stooges in Orbit*. When Larry, Curly, Joe, and Moe meet Professor Danforth, a stereotypical disheveled mad scientist, they help him develop his latest invention: a submarine with treads and a propeller, designed for battle underwater, on land, in the air, and as a bonus, in orbit (thanks to the handy "orbit" button). It's designed for the Department of Defense so the country would be prepared for a Martian attack. Martian spies, however, were already scoping it out in the professor's home. These

Three Stooges brand of aliens look like a cross between Frankenstein's monster and the ugly doctors in *The Twilight Zone's* classic "Eye of the Beholder"—all wearing capes that Dracula would envy.

Initially the Martian leaders watching from the Red Planet simply wanted to invade Earth to gain more living space. But after catching a glimpse of televised dance crazes, traffic jams, war, and a commercial for Polka Dot toothpaste ("for the smile of glamour and beauty!"), the powers that be quickly change their minds.

"If this is Earth, we don't want it! Invasion cancelled!" the leader orders. "Destroy the miserable planet instead."

The spies ultimately gain control of the submarine-tank-helicopter-rocket and mount

a ray gun on top, which disintegrates whatever they aim it at. Spoiler alert: the planet is saved by the Stooges, who had managed to bumble their way aboard before the ship took off and fend off the evil Martians.

Long before this victory, however, Moe, Larry, Curly Joe, and their producer/manager Norman Maurer had to thwart the U.S. government's attacks on the script. A request to film at Oxnard Air Force Base had been denied months before filming because officials didn't appreciate the jokes and gags, which had the potential of making military personnel look bad or stupid. In a letter dated March 30, 1962, Donald E. Baruch, head of the audio-visual department, explained that the Department of Air Force and the Office of the Assistant Secretary of Defense for Public Affairs had reviewed the script and wasn't keen on helping.

"The physical cooperation requested falls within the 'limited cooperation' category," Baruch wrote. "To qualify for this assistance, the policy states that the picture is in the best interests of the Department of Defense and the public good. THE THREE STOOGES MEET THE MARTIANS hardly fits this bill." (His caps, not mine.)

Maurer eventually made several requested script changes to appease the officials and was able to get them to reconsider. He did, however, fight to keep one questionable scene involving a series of pies getting tossed into the faces of Air Force officials.

Maurer argued, "This scene is needed since the Air Police need to be temporarily blinded and because it is extremely funny."

Martian humor went from slapstick to all-out camp in 1964's *Santa Claus Conquers the Martians*. Shot in just ten days, the film features cobbled-together Martians that are nothing more than men with green face paint wearing green clothes and sporting helmets equipped with extendable antennae that they apparently yanked off their TV sets. Unfortunately for them, those TV sets still worked and mesmerized their children with Earth shows. Martian kids watched it closely and gloomily, creating a planet-wide couch potato epidemic. With their future at risk, Kimar, leader of Mars, seeks the help of the wisest Martian alive—an 800-year-old hermit who looks like a cross between Yoda and Dumbledore. Who could be wiser?

The problem is one that Yodumbledore had seen coming: Martian children have information fed into their brains from the day they're born and never learn to have fun. So, now, as they watch the Christmas season unfold on television, they're rebelling. Martians need a Santa Claus of their own to teach their children joy, and there's only place to get one.

That means Kimar and several of his council chiefs head straight to Earth. Their mission is more challenging than expected because, as they discover through their inflight telescopes, there are Santas on every street corner. Which is the real one?

As luck would have it, they stumble upon young Betty and Billy Foster sitting in the woods by themselves, wondering what Martians look like. Little did the curious kids know that their question would be answered immediately. But Betty needs clarification. "Are you a television set?"

"Stupid question!" retorts Voldar, the cranky Martian, before addressing his comrades. "Is this what you want to do to our children on Mars? Turn them into nincompoops like these?"

Betty and Billy, being the nincompoops Voldar said they were, immediately give away the real Santa's location and jeopardize Christmas for everyone. The Martians hop right back in their ship and zip straight up to the North Pole to get their prize. With their trusty freeze rays (painted Wham-O Air Blasters), they ward off the elves and Mrs. Claus and take Santa back home.

IF NOT MARS, WHERE DID LITTLE GREEN MEN COME FROM?

When the world's leading scientists believed in Martians, they imagined them standing up to twenty feet tall and boasting massive chests. Based on the planet's low-gravity environment and thin atmosphere, it was a reasonable guess. Yet ask anyone from the past seventy years what they think of when they think of Martians, and they'll talk about "little green men." But why?

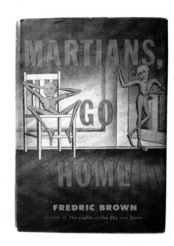

There's no specific genesis to the phrase. Rather, it seems to have been adopted as a term for aliens from earlier uses referencing magical creatures. In the early 1800s, for example, *Sketches in Ireland* referred to leprechauns as "little green men" who rode "cattle not bigger than cats, waving their hunting caps over their heads." Later in the century, fairy tales referred to "little green men" as actual fairies, or in at least one case as tiny folks living in a forest who used sunflowers for tables, spider webs for tablecloths, and cutlery made from stamens and pistils of flowers. Science fiction got its first dose of little green men in a 1946 short story published in *Weird Tales* magazine. In "Mayaya's Little Green Men," by Harold Lawlor, the title creatures mysteriously help Mayaya clean and care for a family without being seen. Two years later a little man named Marvin the Martian was introduced to the world. He wasn't green, but his wardrobe was.

By the 1950s, with the UFO craze in full swing, the press began using the "little green men" expression to describe the creatures who might be scoping out the Earth. A 1952 article even claimed that a pilot passing a flying saucer spotted "little green men staring out of the windows." However, they were believed to be from Venus.

In Fredric Brown's 1955 book *Martians, Go Home*, billions of annoying little green men from Mars "kwim" to Earth (their term for teleportation) and call everyone "Mack" and "Toots." A decade later the Great Gazoo appeared on *The Flintstones*, but instead of "Mack" and "Toots," he addresses stone-agers as "Dum dums." Gazoo wasn't identified as a Martian, but the little green man certainly helped perpetuate the stereotype. Now we're stuck with the moniker, until we discover some form of actual Martian life that requires an entirely different description.

With more than ninety percent of American households owning televisions in the 1960s, the Martian invasion wouldn't have been complete without raiding the country's living rooms. The most notable intruder was Ray Walston, star of *My Favorite Martian*, along with co-star Bill Bixby. The show debuted in 1963 and was spun right out of all the flying saucer hoopla that had dominated headlines over the previous decade. Walston, as the Martian, loses control of his saucer when an Air Force pilot gets in his way and causes him to crash right near a military base. On the bright side, instead of getting scooped up by secretive government officials, he's found by Bixby's newspaperman, Tim O'Hara.

"I could give you some wild tale of being a Russian astronaut or the designer of an experimental spacecraft, but the plain fact of the matter is I'm from Mars," the Martian explains to the confused reporter in the series pilot.

Marooned until he can fix his spaceship, which may require items that "haven't been invented yet," the Martian moves in with O'Hara and is introduced to others as Uncle Martin. It's an easy disguise since Uncle Martin looks like the average avuncular human, except for when his antennae rise and render him invisible. Martians, he explains, use all of their brains instead of the measly portion humans use. That means he has all the powers people would expect from such an advanced being: telepathy; mindreading and the ability to influence minds; levitation with the point of a finger; and not only does he speak perfect English, but he can even speak to animals. All that, and not a single urge to annihilate Earth or steal its women. How could he not be everyone's favorite Martian?

MARTIANS STEAL THE SECRETS
OF OIL FROM EARTH

In *Destination Earth*, Mars is populated by short yellow men (not a woman in sight) with pointy ears, who are ruled by a dictator called Ogg the Magnificent. Produced in 1956 by animator John Sutherland, who once employed William Hanna and Joseph Barbera, the Martians look like intergalactic descendants of the Flintstones.

In need of a better way to power his state limousine, the great and powerful Ogg sends a volunteer to Earth to discover how humans run their vehicles. Amazingly, the Martians can fly through space despite their difficulties driving down streets.

The brave explorer observes the wonders of American transportation and discovers it's all made possible by oil and free-market competition. He hurries back to Mars, announces his findings, and inspires his fellow Martians to start drilling for oil.

The Cold War propaganda cartoon was commissioned by— who else?—the American Petroleum Institute.

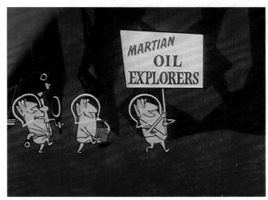

THAT'S THE MARTIAN SIGNPOST UP AHEAD—YOUR NEXT STOP, THE TWILIGHT ZONE

"There is a fifth dimension, beyond that which is known to man. It is a dimension as vast as space and as timeless as infinity." Rod Serling opened *The Twilight Zone* with these words, and in that fifth dimension, aliens often served as a vehicle to explore allegories of human emotions and foibles. Martians, specifically, show up in a secluded diner, a neighborhood bar, and, of course, on the Red Planet itself. In each of these three episodes, Serling uses the men from Mars to remind us that we can all be better Earthlings.

"Will the Real Martian Please Stand Up?"

Bad weather causes a group of stranded bus passengers to wait out the storm at a diner. As if the delay isn't bad enough, they learn that a Martian is among them. A series of unusual events unfolds, including a randomly ringing telephone and a jukebox that plays on its own. Accusations fly and suspicion grows, but finally the weather clears and the paranoid passengers leave. The Martian then returns to the diner and, with a third arm emerging from beneath his sport jacket, informs the chef that he's chosen this site for colonization. The chef, however, is unfazed. He removes his hat to reveal a third eye on his forehead and explains that his people, the Venusians, have beaten them to it.

"People Are Alike All Over"

An astronaut travels 35 million miles to Mars and discovers the people there look just like humans (dressed with a bit of ancient Greek toga chic). Not only do they look human, but they're perfectly polite and provide him with living quarters exactly as he'd enjoy on Earth. Every detail—right down to the rotary phone—was discovered by their ability to peer into his mind. All seems ideal until our Earthling discovers that his habitat is a zoo enclosure. Filled with greed and self-interest, the Martians show us that, like the title proposes, people are indeed alike all over.

"Mr. Dingle, The Strong"

A two-headed Martian with elongated foreheads, antennae, and a boxy mechanical torso visits Earth and heads straight to a bar. Not for a drink, but to give an old, feeble man, Mr. Dingle, superhuman strength to see what he does with it. Would he perform heroic feats and help society? Of course not. When Mr. Dingle uses it for nothing more than exhibitionism, they strip his powers and move on to another planet. Fortunately for Mr. Dingle, a pair of Venusians arrive right after and give him another shot—this time with super-intelligence.

THE WRITERS WHO REALLY GROKKED MARS

When the first *Viking* landed on Mars in 1976, Ray Bradbury was invited to witness the event at NASA Jet Propulsion Laboratory. The author of *The Martian Chronicles* saw his imagination become reality right before his eyes. As the robot fed images back to Earth, an interviewer asked Bradbury where the Martians were. "Don't be a fool. WE are the Martians!" he responded. "We're going to be here for the next million years. At long last, WE ARE MARTIANS!"

It's what he'd been saying since *Chronicles'* publication in 1950, when the last human settlers of his seminal book gaze at their reflections in the canals and discover that Martians look just like us.[11]

Sooner or later, Bradbury will be right, but in his tale, the planet was populated before people from Earth arrived. Humanoid in form and advanced in intelligence, they lived in crystal houses surrounded by canals, read books that sang, ate golden fruits that grew on crystal walls, and communicated telepathically. Then, waves of rockets from Earth arrive and everything good is spoiled. Like the Europeans arriving on North America's shores in the sixteenth century, the visitors bring their germs and waste no time spreading them. Within a few years, nearly all the Martians are wiped out by chicken pox.

One of Bradbury's early explorers mourns the loss of Martian life and fears for the complete elimination of the culture they've left behind. "We Earth men have a talent for ruining big, beautiful things," he says. "The only reason we didn't set up hot dog stands in the midst of the Egyptian temple of Karnak is because it was out of the way and served no large commercial purpose."

---

11 Bradbury created *The Martian Chronicles* by weaving together a series of his Mars-related short stories at the suggestion of his editor. The book's action originally began in 1999; however, when that distant tomorrow became not so distant in 1997, subsequent printings pushed the futuristic tale to 2030 for modern-day readers.

Sure enough, as the dystopian society develops, an entrepreneur sets up a hot dog stand awaiting a flood of rockets from Earth. As New New York, New Chicago, and other cities develop, life on Mars becomes filled with fear, hatred, and censorship.

Those same destructive tendencies, having been cultivated and spread for thousands of years back on Earth, eventually lead to a devastating world war. It's a moment that reflects the political climate Bradbury was writing in but still rings true today. Light signals flash in the sky with a Morse code message to Mars, urging settlers to return home. With a dying planet, screaming across space for help, Bradbury flipped Percival Lowell's script from earlier in the century. All but a handful of colonists return, leaving only a few on Mars to keep whatever is left of humanity alive.

A year after *The Martian Chronicles* swept the Earth, another young writer found inspiration in the Red Planet for his first novel. Though Arthur C. Clarke would later gain widespread fame as the author of *2001: A Space Odyssey*, at the time he was known as a science journalist and chairman of the British Interplanetary Society. Those interests led to *Sands of Mars*, the story of a writer named Martin Gibson who journeys to Mars to visit its developing colony. Upon his arrival he meets with the Martian settlement's Chief Executive and learns of their terraforming efforts, along with the various political and cultural issues the colonists face. Earth, for example, is wondering what good Mars is doing for everyday people back home. What are they getting in return? Is it worth the costs? And from the Martian colonists' perspective, why should they take directions from politicians millions of miles away?

Gibson soon discovers that they have a plan in the works to help answer these concerns while achieving a greater degree of independence and quickening the terraforming process. All they have to do is set Phobos on fire. As Clarke puts it, they've lost a moon but gained a sun that will burn for a thousand years. Mars will warm up and build an atmosphere, meaning that as the generations of Martians continue, they'll have less need of Earth's financial and material support.

Clarke, who may have seen himself as Gibson, wasn't just creating a science-fiction novel, he was predicting humanity's eventual colonization of Mars. The author often proved himself to be a successful futurist. In 1964, for example, he suggested the idea of Internet search engines and cell phone technology—

"No colonist or explorer setting sail from his native land ever left so much behind as I am leaving now. Down beneath those clouds lies the whole of human history; soon I shall be able to eclipse with my little finger what was, until a lifetime ago, all of Man's dominion and everything that his art had saved from time."

—Martin Gibson
looking back toward Earth on his way to Mars
in Arthur C. Clarke's *Sands of Mars*, 1951.

including video chat—that would create "a world where we can be in instant contact with each other wherever we may be."

Although Martian settlements have not yet come to pass, Clarke wrote *Sands of Mars* ten years before people first flew to space. So, he got that part right, even though the entire sci-fi story surely sounded more like fiction than science. Of course, that was his goal with any prediction.

"The only thing we can be sure of in the future is that it will be absolutely fantastic. So if what I say now seems to you to be very reasonable, then I will have failed completely," he once said. "Only if what I tell you appears absolutely unbelievable have we any chance of visualizing the future as it really will happen."

Today, Clarke's colonization ideas are being discussed as eventual realities, except maybe for the part about turning Phobos into a ball of fire. Then again, it's a perfect example of an "absolutely unbelievable" prediction.

Robert A. Heinlein's 1961 masterpiece *Stranger in a Strange Land* (*right*) tells the story of Valentine Michael Smith, a human born on Mars and raised by Martians before being returned to Earth as an adult. The novel takes place sometime in the future, when new laws dictate that Smith, the Man from Mars, has inherited a fortune and becomes the sole owner of the Red Planet. Of course, being new to Earth and trapped in a hospital, he doesn't know about any of this.

In addition to money, Smith finds himself with a nurse and a lawyer on his side. The two befriend him by sharing water, becoming his "water brothers." On Mars, we learn, sharing water brings closeness between two people. These new water brothers help him escape the clutches of government shenanigans and legal entanglements, leaving him free and rich.

With everything being completely new to the Man from Mars, Heinlein's character gives him a means to examine the world we call home, as well as its many peculiarities—to grok all the nuances of society. In fact, so much grokking was done to make sense of life on Earth that the author's Martian term became an actual English word, defined by Merriam-Webster as "to understand profoundly and intuitively."

For Heinlein, the "two biggest, fattest sacred cows" to grok in Western culture were sex and God. It doesn't take long for Smith to discover that the former allows humans to

get even closer than sharing water. He finds it to his liking. Religion, on the other hand, is a confusing notion. His water brothers take him to visit the Supreme Bishop of a widespread religious group, but Smith's notion of God is far different from the version they preach. Instead, he sees each and every person as having God within them, explaining to all who will listen, "Thou art God."

As the Man from Mars ventures out into this strange world so far from home, he takes up odd jobs, including a stint as a magician at a traveling carnival. Smith's Martian powers of telekinesis allow him to levitate people on stage, make juggling balls disappear, and perform other feats of actual magic, but no one believes it. He lacks showmanship and is eventually fired. Yet the experience teaches him valuable lessons about the human psyche, particularly how to draw marks in, turn the tip, and sell the blow-off (or, in non-carny lingo: how to attract a crowd, make them pay, and then make them pay again for one final act). It's the same sort of scheme the religious group used to build its following, only with a different kind of show.

Smith realizes he can create his own religion that merges human needs with his altruistic Martian ideals. The Church of All Worlds, as he calls it, becomes a cult where people find true happiness in communal living spaces called Nests. No possessions or clothes are needed—not as long as they've got group sex and reminders that God is within each of them.

Heinlein had experienced the cult mentality decades earlier while attending parties with L. Ron Hubbard at Jack Parsons's OTO house. Perhaps recalling those cult experiences, Hein-lein includes a Crowley text as part of Smith's self-education.

Within a year of the book's publication, the ideas expounded by the Church of All Worlds had inspired a neo-pagan cult of the same name. Its leader, a Missouri college student named Timothy Zell, changed his own name to the more magical Oberon Zell-Ravenheart and moved his growing flock to California. By 1968 the religion became legally recognized; it promoted polygamy, socialism, and the notion that the only belief should be a "lack of beliefs."

Heinlein abstained from joining, but he did subscribe to the CAW newsletter, the "Green Egg," to keep up with their activities. One story he may have read about in 1980 was their mission to create unicorns by fusing a goat's horns into a single one. This odd experiment brought the cult full circle with Valentine Michael Smith's carny days, as several of these "unicorns" went on tour with the Ringling Bros. and Barnum & Bailey Circus. Though its owners probably didn't realize it, in a circuitous sort of way, the Greatest Show on Earth had an act from Mars.

"FOR THE MARS THAT USED TO BE, BUT NEVER WAS."

Arthur Bertram Chandler used those words in the dedication of his 1965 science-fiction novel, *The Alternate Martians*. A year earlier, *Mariner 4* neared the surface of Mars and revealed a barren planet far from the visions of Edgar Rice Burroughs, H. G. Wells, and others. There was no breathable atmosphere, no climate warm enough for scantily clad princesses, and not a single canal to sail upon through the Martian landscape. But Chandler wasn't ready to give up on their adventurous imaginations.

The Alternate Martians stars a group of scientists with a love for science fiction and a time machine that lets them explore alternate universes. With all that power, they choose to visit Mars to "prove that some fantastically bad guesses made by twentieth century science fiction writers hadn't been bad guesses at all, but faulty memories." Sure enough, they encounter lesser versions of the characters from Burroughs' Barsoom series (with cockney accents) and evil octopus-like blobs from *The War of the Worlds*. The early writers hadn't guessed wrong at all, Chandler proposes; they just embellished their memories of alternate universes for the sake of better stories.

Shortly after Michael Valentine Smith journeyed to the Third Rock, so did Thomas Jerome Newton. In *The Man Who Fell to Earth*, author Walter Tevis, like Bradbury, uses Mars as a way to warn us about the danger of humankind destroying itself here on Earth. An atomic war on the Red Planet, or Anthea, as Newton calls it, left only a few survivors. So before they all die out, he used the last bit of Anthean fuel and took a one-man spacecraft straight to Earth. More specifically: Kentucky. There, he planned to find fresh resources and build new spaceships to bring his people to join him.

Newton, who is humanoid in form, blends in despite his hardly southern accent and uses his advanced intelligence to offer the world fantastic new inventions, most notably sharper film that develops itself. With the help of a patent lawyer, he soon becomes a multimillionaire and assembles a team to begin building a space vehicle. His dedicated workers are led to believe its purpose is merely for exploration.

Before falling to Earth, Newton spent fifteen years honing his plan watching American, British, and Russian television broadcasts to learn the languages, mannerisms, geography, and history of humankind. Along the way, he discovers there is much more to humanity than what's shown on TV, and it's not always pretty. For a while, he remains hopeful that he'll save his fellow Antheans and together, they can help their new planetary hosts save themselves. As he explains to one of his employees who learns of his true alien nature:

> *"It dismays us greatly to see what you are about to do with such a beautiful, fertile world. We destroyed ours a long time ago, but we had so much less to begin with than you have here. Do you realize that you will not only wreck your civilization, such as it is, and kill most of your people; but that you will also poison the fish in your rivers, the squirrels in your trees, the flocks of birds, the soil, the water? There are times when you seem, to us, like apes loose in a museum, carrying knives, slashing the canvases, breaking the statuary with hammers."*

Over time, however, his hope is abandoned as he succumbs to alcoholism and loneliness. In that sense, Newton becomes more human than his plan ever required.

DAVID BOWIE FELL TO EARTH

David Bowie's hit songs "Space Oddity" (1969, released just five days before *Apollo 11* landed on the moon), "Life on Mars?" (1971) and "Starman" from the art-rock album *The Rise and Fall of Ziggy Stardust and the Spiders from Mars* (1972) made it clear that the musician had a penchant for space. If any rock star could've toured on the Red Planet, it would've been him. Instead, Bowie took on a new persona and became a Martian, on film, starring as Thomas Jerome Newton in Nicolas Roeg's 1976 big screen adaptation of *The Man Who Fell to Earth*.

A generation later Bowie's space-themed music made it there, thanks to the astronauts on the space shuttle *Columbia*, who woke up to "Space Oddity" during their 1996 journey, and to Canadian astronaut Chris Hadfield, who in 2013 played an acoustic version of the song from the International Space Station. Millions viewed his rendition on YouTube, including David Bowie. "He described it as the most poignant version of the song ever done, which just floored me," Hadfield later said. After Bowie's passing in 2016, the astronaut tweeted: "Ashes to ashes, dust to stardust. Your brilliance inspired us all. Goodbye Starman."

Elon Musk kept Bowie's love of the cosmos alive by playing "Life on Mars?" during the February 6, 2018, launch of a SpaceX rocket carrying a Tesla Roadster. The car was aimed toward Mars, playing Bowie's "Starman" on a loop. Ziggy Stardust may finally discover what life on Mars is like, or at least life near Mars.

Science fiction's dire warnings of self-destruction simmered down after the Cold War ended in 1991. Thankfully, it looked like we weren't going to blow ourselves to smithereens and end up a dead planet like Mars. In fact, now that the world felt a little safer and we were decades past the sci-fi B-movie era and beliefs in intelligent life on Mars, we could finally make fun ourselves. *Mars Attacks!* did just that.

Tim Burton's 1996 spoof opens with a massive fleet of flying saucers ominously approaching Earth. Rather than fear the worst, America's preeminent scientist assures whoever will listen that the Martians are clearly an advanced civilization, for they had mastered interplanetary travel and, therefore, would be well past the concept of warlike aggression. In a televised interview, he explains that there's a good reason our space program hadn't yet discovered the Martians. "We didn't get into the canals," he tells the audience. "The Martian canals are actually canyons. Some of them are over a hundred miles deep. Martian civilization was clearly developed under the surface of the planet. The science and technology must be absolutely mind boggling."

Burton's Martians, as we soon discover through a video transmission, are the proverbial little green men with rail-thin limbs and giant heads (more specifically, giant exposed brains protected by glass bubbles). In his studies of the creatures, the scientist proposes that the "large cerebrum indicates telepathic potential." Every description evokes the many Martian theories offered by early thinkers and writers, while the country's optimism and anticipation reflect the public's fervor of the early twentieth century. It's all very exciting for everyone until it's revealed that the Martians are much closer to those imagined by H. G. Wells. The events that follow harken back to the panic created by the 1938 broadcast of *The War of the Worlds*.

MARS ATTACKS—WITH BUBBLE GUM!

When *Mars Attacks!* first descended on our planet, every Martian came armed with a pink stick of gum. Though today it's best known as a Tim Burton film, the filmmaker based it on the 1962 Topps Chewing Gum sci-fi trading card series of the same name. The brainy Martians were inspired by a 1950s *Weird Science* comic book and the creatures from 1955's *This Island Earth*. Each of the collectible cards depicted brutal battle scenes erupting all across Earth—mostly with the Martians ravaging our world with their deadly disintegrator weapons. Comic book and sci-fi fans surely loved them, but public outrage over the graphic illustrations of gore and destruction forced the company to halt production.

Nine days before the release of *Mars Attacks!* NASA launched *Pathfinder* and dreams of sending humans to Mars once again began to look like an eventuality. Ever since *Viking* shared photos of the Martian surface and provided reams of data, writers have been able to envision colonization with much greater specificity and plausibility than the stories of their predecessors. Kim Stanley Robinson spent a decade researching papers on the Red Planet and theories on terraforming before publishing *Red Mars* in 1993, the first of his Mars trilogy. The saga begins in 2026, when a group of Americans, Russians, and international explorers dubbed the First Hundred venture to Mars to begin a new society. With the aid of robotics, natural resources, and their diverse expertise, they successfully build cities and cultivate food. There are no native Martian creatures or Lowellian canals to contend with, just humanity, which offers its own set of ethical and social challenges. Factions begin to form, namely those who wish to terraform Mars with orbital mirrors and deep cylindrical excavations that heat the atmosphere, and those who long to keep it pure. The struggle stretches beyond the First Hundred as the years pass and a million new settlers from different countries and cultures join them. Earth's own interest in using Mars for its resources doesn't help either. The new society begins to reflect the old as a revolution—"some kind of sci-fi 1776"—ultimately brings death and destruction. Robinson, much like Bradbury, reminds us that when we finally escape Earth, we'll bring our baggage with us.

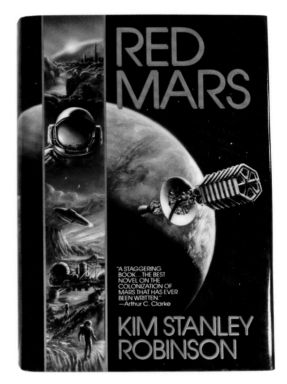

"The beauty of Mars exists in the human mind. Without the human presence it is just a random collection of atoms, no different than any other speck of matter in the universe. It's we who understand it, and we who give it meaning. All our centuries of looking up at the night sky and watching it wander through the stars. All those nights of watching it through the telescopes, looking at a tiny disk trying to see canals in the albedo changes. All those dumb sci-fi novels with their monsters and their maidens and dying civilizations. And the scientists who studied the data, or got us here. That's what makes Mars beautiful. Not the basalt and the oxides. ... We can transform Mars and build it like you would build a cathedral, as a monument to humanity and the universe both."

—Sax Russell,
   arguing for the need to terraform
   in Kim Stanley Robinson's *Red Mars*

"We are not the lords of the universe. We're one small part of it. We may be its consciousness, but being the consciousness of the universe does not mean turning it all into a mirror image of us."

—Ann Clayborne,
   in her rebuttal, *Red Mars*

Software engineer Andy Weir further explored the idea of humans on Mars—or more specifically, a *human* on Mars—in his 2011 novel *The Martian*. Set in 2035, the story features a protagonist named Mark Watney, who is a botanist stranded on the Red Planet after a mission goes awry. Trapped with limited supplies, he manages to use every available resource to survive—including his own poop as fertilizer for growing potatoes—and establishes communications with Earth by finding the *Pathfinder* lander and successfully rebooting it.

Writing nearly two decades after *Red Mars*, Weir took advantage of knowledge gathered from years of planetary research and data to inform his narrative. He even wrote his own software to calculate the orbital paths required for the story. *The Martian*—including Ridley Scott's 2015 adaptation (*above*)—feels like it could become a true story sometime within the next few decades. A reconnaissance trip for the First Hundred, if you will. After more than a century of tales filled with Martian battles and invasions, Weir and Robinson have helped prepare the science fiction genre to evolve into what many early writers knew and hoped would happen eventually: become science fact.

ANDY WEIR ON FINDING LIFE ON MARS

"I think Mars doesn't have any life at all and I think it never has. My reasoning is this: Let's say you grabbed a sample from anywhere on Earth. Anywhere at all. The air. The ocean. The sands of the Sahara Desert. The icy slopes of Antarctica. Anywhere at all. You could grab a teeny, tiny sample from any of those areas, look at it under a microscope, and it would be TEEMING with life. Microbes everywhere—even places that don't have obvious macroscopic life.

"Evolution is incredibly good at finding solutions to problems and working out how to live in any environment. So I believe that if life evolves anywhere on a planet, it will spread and evolve and differentiate until it's *everywhere* on the planet. And so far, we have never found any life of any kind on Mars. To me, that means there's never been any in the first place.

"There are theories that there was once life, which then died off or went underground as the water went away, but I don't buy that either. The erosion of Mars's atmosphere and loss of its oceans happened over hundreds of millions of years. Evolution can easily handle adapting to an environment that's changing so slowly."

CHAPTER 6
LEAVING THE CRADLE

Settle on Mars, one of the first heavenly bodies to cool off after the creation. We have practically no water, practically no oxygen in our atmosphere. The oxygen has gone into the rocks, and much of our country is covered with fine red dust, caused by the oxidation of our iron deposits.

"The dust doesn't always stay put, however, because we have winds of gale force almost all the time, caused by rapid changes in temperature, which ranges from 30 to 80 degrees at noontime but falls to 100 degrees below zero at night.

"We have no twilight and no dawn. Our daytime sky is no brighter than twilight on Earth, but there's every chance of romance here. We have two nice moons, one of which remains 132 hours above the horizon. Our nearer moon, Phobos, is as close to us as New York is to Los Angeles on Earth. And if you like eclipses, they are a dime a dozen up here. We have as many as 1,000 of them each year.

"Come, spend a year with us. You'll have plenty of time to look around, because our year has 687 days. And girls—if you happen to weigh about 200 pounds, you'll find life can be beautiful on Mars, because here you'll weigh less than 80."

This was how a reporter imagined the Mars Chamber of Commerce might entice visitors in 1939. Unfortunately, the pitch missed a few highlights for the outdoorsy types, such as Olympus Mons, the sixteen-mile-high mountain that's as big as Arizona, and Valles Marineris, the solar system's largest canyon. At 2,500 miles long, it makes the Grand Canyon seem not so grand. If climbing's not your thing, there's the massive ice-filled crater that runs more than a mile deep near the north pole, found in 2018 by the European Space Agency's Mars Express probe. The ice skating promises to be spectacular as long as you can glide and spin in a bulky spacesuit. At the time, the idea of visiting Mars existed only in the realm of fantasy. Today, the concept of going there and even establishing colonies is very real.

Despite its geological wonders, who would want to go to such a cold, dead place? Well, a lot of people.

SUMMIT

OLYMPUS MONS

The solar system's highest peak

According to Mars One, an organization planning a human settlement on the Red Planet, there are at least two hundred thousand brave, thrill-seeking, history-chasing, or just plain suicidal people anxious to make a one-way trip. The group says it received that many applications in its 2013 call for volunteers and has so far narrowed the pool to one hundred candidates. After it builds an outpost with rovers in the 2020s, Mars One plans to send its first crew of humans by 2031.

Enthusiasm is widespread for various reasons. Take Explore Mars, for example, a nonprofit that runs an annual Humans to Mars Summit in Washington DC to bring together the aerospace community for three days of collaborating, educating, and networking. Its leadership is eager to push for goals in the space program, boost public morale for exploration, and continue building on what we've already learned about our planetary neighbor. They hope to put a bootprint on Mars by 2033. "That is a particularly good year for planetary alignment," says CEO and cofounder Chris Carberry. "You can get a lot done in that specific year. The planets align, so it doesn't take as much energy to get there."

In April 2019, NASA administrator Jim Bridenstine aligned the administraton's plans with Explore Mars's 2033 goal during a congressional hearing with the House Committee on Science, Space, and Technology. NASA will use a 2024 moon mission as a proving ground to test various capabilities and technologies on another world. Holding a trial run only three days away from Earth helps, if help is needed. The upcoming mission assumes that NASA's rocket, the giant Space Launch System (SLS), will be ready in time.

Dr. Robert Zubrin, aerospace engineer and founder of the Mars Society, has been ready to fly to and colonize the Red Planet since the mid-1990s, when he developed Mars Direct—a plan to reach the planet using its natural resources and existing technology. Thorough in its details, Zubrin's plan included mining water, using raw materials to produce metals, plastics, glass, and ceramics, developing agriculture, and even creating fish farms to grow tilapia so colonists wouldn't be forced into vegetarianism.

Then there's Buzz Aldrin, the second man to walk on the moon who would gladly be the first to walk on Mars. He'd love to see Zubrin's plan come to fruition by 2035. Why two years after Carberry's goal? Perhaps to take the time to ensure we're fully prepared, though Aldrin also points out a poetic symmetry to the date: it's 66 years after the moon landing, which was 66 years after the Wright Brothers first took flight.

To make it happen, he's developed his own scheme that offers a few differences from the Mars Direct approach, namely the use of cyclers to get travelers, cargo, and any other necessary materials to and from the Red Planet efficiently. Essentially, these spacecrafts would perpetually cycle back and forth between the planets, saving on launch fuel and the need for new rockets with every journey. A network of space taxis would take passengers to the cyclers when they reach Earth's orbit and pick them up at Mars to deliver them to the surface. Like hotel shuttle buses at the airport.

"It's a mix of beautiful simplicity melded with a ballet of gravitational forces that moves humanity outward to Mars," Aldrin wrote in his 2013 book *Mission to Mars*.

The famed astronaut also advocates for reaching Phobos first, allowing humans to establish a local office from which we can better monitor and control robots on Mars as they build habitat modules and other infrastructure. Once complete, we can safely land a crew.

Of course, getting humans to Mars doesn't have to mean living there. Geologists, for example, would benefit from an exploratory trip to learn more about the planet significantly faster. According to an estimate from Steve Squyres, it would take a human thirty to forty-five seconds to do what a rover does in one full day. Pascal Lee, director of the Mars Institute, imagines setting up outposts on the surface for scientific research, with crews rotating in and out, much like what's done at Antarctica.

Before his passing in 2017, Dr. Stephen Hawking challenged humanity to get to the Red Planet, not just for science or reaching extraordinary new milestones, but as insurance for the survival of the species. He believed that we were outgrowing Earth and damaging it by draining its resources and causing climate change.

"When we have reached similar crises in our history there has usually been somewhere else to colonize," Hawking wrote. "Columbus did it in 1492 when he discovered the new world. But now there is no new world. No Utopia around the corner. We are running out of space and the only places to go to are other worlds."

To make matters worse, Hawking warned against the threat of a giant asteroid hurtling through space and striking the Earth. He feared we'd be defenseless against such a random, inadvertent collision. If we don't populate another place in the solar system, we'll put ourselves at risk of getting wiped out.

"The last big such collision with us was about sixty-six million years

"For me, getting humans to Mars is definitely not just footsteps and flags, and it's not just doing research. For me, it is settling. It's to start a new branch of civilization, which I fervently hope will be more civilized than sometimes we are here on Earth. Not because we'll be better humans, but because we'll simply need each other, which will probably make us a bit more friendly to each other regardless of who we are or what our job title is on Mars. There is a chance that we will finally appreciate each other for the contribution we all bring to the table to keep us healthy, alive and happy. That's my driving motto: Try to make a better society."

—Artemis Westenberg,
President Emeritus and Cofounder
of Explore Mars, in 2019

ago and that is thought to have killed the dinosaurs, and it will happen again," Hawking claimed. "This is not science fiction; it is guaranteed by the laws of physics and probability."

His timeline for escape: fifty years.

That's a lot of doom and gloom. No one wants to be annihilated like the dinosaurs, but if we want to set up shop on another planet, we may need an even greater motivation for our governments to fork over the kind of cash it would take to expedite the mission. Specifically, political motivation. That's how President John F. Kennedy got us to the moon. When he challenged NASA to plant a flag on the lunar surface by the end of the 1960s, it wasn't just to learn more about the moon—it was to beat Russia and flex American might and intelligence.

"The number one reason to send people into space is to somehow maintain or acquire a competitive edge in the geopolitics of it all," Pascal Lee says. "Once you've committed humans to go, what else is there to do in space but science? Obviously not to open a law firm or dental office."

So who can give America an excuse to put people on Mars? Lee thinks China might. "Then it's the whole national geopolitics thing kicking in; it'd become imperative for us to go there."

Reaching this newest frontier would be a historic achievement for humanity and a testament to our ingenuity, resolve, and perseverance. Just as importantly, if not more importantly, it would inspire future generations to continue the progress. Aldrin should know better than anyone, since his Apollo mission to the moon did exactly that for many of today's NASA engineers—not to mention a few ambitious entrepreneurs who can operate outside of geopolitics and need nothing more than their own personal motivations.

THE RACE OF THE BILLIONAIRES

Sure, getting to Mars will require extraordinary feats of engineering, but it could also use a fresh dose of good old-fashioned human curiosity, innovation, and business acumen. A trio of billionaires in particular is oozing with all three: Jeff Bezos, Richard Branson, and Elon Musk.

Bezos, whose world domination is in full swing with Amazon two-day shipping and promises of delivering toothpaste via drones, is ready to expand from Earth and conquer the universe. The space bug bit him as a kid during reruns of *Star Trek* and thoroughly planted the seed that's grown him into a solar system entrepreneur. Bezos's pursuits continued in college at Princeton, where he became president of the campus chapter of Students for the Exploration and Development of Space. By 2000 he had started turning his goals into reality by founding Blue Origin, a company that's determined to transform space into the New Frontier. But he's not driven simply by fascination and wonder. Bezos believes an energy crisis will hit within a hundred years and limit progress on Earth. To avoid the catastrophe, he wants people to live and work in space, with Mars being just one possible destination. After all, there's plenty of space in space, and lots of asteroids to populate.

"The solar system can easily support a trillion humans," Bezos has said. "And if we had a trillion humans, we would have a thousand Einsteins and a thousand Mozarts and unlimited, for all practical purposes, resources and solar power unlimited for all practical purposes. That's the world that I want my great-grandchildren's great-grandchildren to live in."

If Bezos has his way, then Thomas Dick's theoretical census of the universe in 1838 may eventually become a necessity. Such a world would open up plenty of job opportunities for those trillion people. Life in space and on other celestial bodies would require advancements and innovations in travel, agriculture, architecture, communications, engineering, geology, chemistry, medicine, nutrition, plumbing, fashion, souvenir snowglobe-making, and, well, just about every industry imaginable plus a few new ones. Those thousand Einsteins will come in handy.

Richard Branson, founder of the Virgin empire, has been striving for commercial space travel since 2004. His company, Virgin Galactic, has already successfully launched a crew fifty miles above the Earth in SpaceShipTwo, becoming the first suborbital flight sprung from U.S. soil since the end of the space shuttle program in 2011. Branson isn't aiming for Mars, but his efforts will undoubtedly help lead the way.

That brings us to Elon Musk. The founder of Tesla who is driven like his electric car's namesake and has the passion and money for Mars like Percival Lowell. In 2002 Musk founded Space Exploration Technologies (SpaceX) with the focused goal of going to Mars and colonizing it. To borrow a term from his car, he's shifted SpaceX into Insane Mode by planning to send two Big Falcon Rockets to the Red Planet by 2022. By "Big," he means big. Each rocket measures 387 feet in length. That's considerably longer than a football field. They'll be loaded with a hundred tons of supplies and materials to begin finding the best sources of water and putting in place an infrastructure for crewed missions. Musk envisions those missions launching two years later to begin building a propellant plant, installing a massive array of solar panels, mining and building a propellant plant, installing a massive array of solar panels, mining and refining water, and performing other necessary tasks to start a base. Once that's done, he'll be ready to populate cities—up to a hundred passengers at a time in his interplanetary spacecraft. Journeys are expected to last three to six months. Like frontiersmen in their covered wagons, early settlers may live aboard the starship until habitats become available.

Like Hawking, Musk believes that finding a home away from Earth is critical. "Are we on a path to becoming a multi-planet species or not? If we're not, well, that's not a very bright future," he has stated. "We'll simply be hang-ing out on Earth until some eventual calamity claims us."

SpaceX may well be the first organization to get humans to Mars, even if its initial plans are nothing more than hype to excite and intrigue the public. But whoever makes it happen will have to overcome a lot of problems. Before we ever set foot on the Martian surface, what will the journey be like? Aside from the boredom that's sure to set in after a few weeks of marveling at the wonders of the universe and gazing at the gradual disappearance of the Earth, two major forces could really put a damper on things: radiation and microgravity.

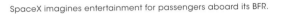

SpaceX imagines entertainment for passengers aboard its BFR.

YOU'VE MADE IT TO MARS! NOW DON'T COME BACK.

The first astronauts to step foot on Mars will have conquered the cosmos and earned eternal bragging rights. That's great, but here's the big question: Will they be able to come home? It's not a matter of rocketry or a freak accident as portrayed in *The Martian*—it's a question of law back on Earth, and a concern for what might come back with them.

This brings us back to *Viking*, the only mission to have conducted tests for life. Since the Labeled Release results cannot be ruled out entirely, scientists cannot be certain that microorganisms, including dormant ones in the dirt, don't exist on Mars. "Are we confident enough that the Martian soil is lifeless to send astronauts? And then to bring those astronauts back to Earth?" asks Chris McKay. "The Outer Space Treaty prohibits 'adverse changes in the environment of the Earth resulting from the introduction of extraterrestrial matter.'"

McKay notes that federal agencies, like the FDA and the Department of Agriculture, could get involved. So could private individuals through the courts if they fear the return mission doesn't comply with the treaty. "Thus, there is the specter of astronauts on Mars prohibited from returning to Earth based on a successful lawsuit claiming a lack of due diligence in the question of Martian life as required by the Outer Space Treaty," McKay says. Proper due diligence would include establishing a lack of harmful entities in the Martian soil. Future robotic missions could help do just that and ensure a roundtrip ticket. Or, if life is confirmed, delay such a trip entirely.

THE SPACE POOP
CHALLENGE

That was the official name given to a 2016 crowdsourced contest to develop easier and safer ways to remain regular in space. In other words, NASA had a shitty job and offered someone a prize to do it. The contest description read as follows:

> The US National Aeronautics and Space Administration (NASA) seeks proposed solutions for fecal, urine, and menstrual management systems to be used in the crew's launch and entry suits over a continuous duration of up to 144 hours. An in-suit waste management system would be beneficial for contingency scenarios or for any long duration tasks.

Let's put this in perspective. It's been more than fifty years since the greatest engineers in the world figured out a way to put a man on the moon. We've since landed multiple robots on Mars and now we're talking about sending humans there, too. Yet by the end of 2016, scientists were still stumped about the best way to take an interplanetary crap.

Normally, the spacecraft provides all the shelter, clean air, water, and food that astronauts need. However, in the event of a major equipment malfunction wherein the cabin loses pressure, the crew would need to change into their space suits. That suit becomes their new shelter, providing them with everything they need until they can return home or resolve the issue. Before the Space Poop Challenge, astronauts wore diapers. Those work fine for eight to twelve hours, but if too much time passes, the waste can cause an infection or sepsis, which can be harmful or even fatal. Death by doo-doo is no way to go.

NASA hasn't sent anyone past low Earth orbit since the last moon mission in 1972, but when astronauts head to Mars, they'll need to safely manage their bodily waste.

Judges reviewed five thousand submissions spanning 130 countries. The winner was Dr. Colonel Thatcher Cardon, who took home a $15,000 prize. His perineal access port (PAP) is a valve opening in the crotch of the spacesuit that would allow the insertion of toilet apparatuses, like an inflatable bedpan, to catch and remove waste.[12]

The implications of this design modification are hugely significant to humankind. If mishandled poop caused an astronaut to die en route to Mars, all of our colonization plans might go straight down the toilet that wasn't there.

---

12 The perineum is the area more commonly and humorously called the taint.

SPACE HATES YOU

No matter how much we love space, the feeling is not mutual. Take microgravity, or zero gravity, for example. Floating around in space may look like fun, but what starts happening to your insides isn't fun at all. The lack of gravity means your muscles don't have to work much anymore, your heart doesn't have to pump as hard, and your bones don't have to support its usual weight. Everything starts to weaken, and your fluids shift about aimlessly. That's not good, especially when it's time to pee—because you don't know when that time will be. Thanks to microgravity, urine doesn't start pushing the bottom of your bladder to tell you you gotta go until your bladder is dangerously full. Those aboard the ISS combat these problems with compression suits, rigorous exercise, and frequent peeing routines to keep their systems as normal as possible, but such practices go only so far.

Then there's that whole radiation issue. Once you're away from the cozy protection of Earth's atmosphere and magnetosphere, your body gets slammed with ionized radiation from every corner of the universe. Every 19.4 hours spent in interplanetary space pummels your body with as much radiation as you are exposed to during a year on Earth. It's like one big cancer shower.

U.S. astronaut Scott Kelly flirted with these dangers and dipped his toes in the waters of long-distance space travel by spending 340 consecutive days aboard the International Space Station in 2015–16. The plan was to find out what a full year in space does to the human body, but Kelly was pulled back early after feeling the effects of the prolonged stay. Upon reuniting with Earth, he soon felt the repercussions as his body reacclimated to gravity. "At every stage I feel like I'm fighting through quicksand," Kelly wrote in his autobiography. "When I'm finally vertical, the pain in my legs is awful, and on top of that pain I feel a sensation that's even more alarming: it feels as though all the blood in my body is rushing to my legs, like the sensation of the blood rushing to your head when you do a handstand, but in reverse." Rashes covered his flesh wherever it made contact with anything it hadn't touched in eleven months—like a chair and a bed.

In addition, prolonged exposure to radiation gives cancer more opportunities later in life than anyone should have. Astronauts soaking in those galactic cosmic rays also face the added risks of cataracts, circulatory diseases, and a messed-up central nervous system. Yet the ISS enjoys some shielding from the magnetosphere. All that said, Kelly still believes that reaching Mars is the next big step and believes we can do it.

Russian cosmonaut and space medicine specialist Valery Polyakov made his own case for space travel in 1994–95, having voluntarily spent 438 consecutive days aboard the *Mir* space station to see if a Mars flight was doable. Upon returning to Earth, his first words were: "We can fly to Mars." Today, at age seventy-eight, he remains healthy.

> ## "SPACE DOESN'T GIVE A DAMN ABOUT IDEOLOGY. IT'S ALWAYS TRYING TO KILL YOU."
>
> **—Dr. Jim Logan,**
> former NASA flight surgeon
> and cofounder of Space
> Enterprise Institute

"So the big change is that you're not going to say, 'Hello Houston, we have a problem,' and let the people on the ground solve the problem for you. You need to have all the tools to solve the problem for yourself because you don't want to wait twenty minutes for them to say, 'Hey I didn't hear that, did that break up?'"

—Dr. Julie Robinson,
chief scientist, ISS, NASA Johnson Space Center, discussing the need for using AI and information technology to make systems more autonomous at the 2019 Humans to Mars Summit

As Elon Musk pushes toward his goal, the aggressive schedule he's set forth only seems feasible because he is less concerned about the physical toll of space on the human body. "There's going to be some risk of radiation, but it's not deadly," he said in 2016. "There will be some slightly increased risk of cancer, but I think it's relatively minor.... The radiation thing is often brought up, but I think it's not too big of a deal."

In addition to its cabin space, storage space, galley, and entertainment area, Musk's BFS (Big Falcon Spacecraft) will offer solar-storm shelter space. Details on how protective that might be are unavailable.

Opinions on whether or not radiation and microgravity are showstoppers depend on whom you ask. NASA reported that the *Curiosity* rover absorbed about half a sievert during its 253-day journey to Mars. (A sievert is how scientists measure radiation dosage.) If it took astronauts six months to travel to Mars and six months to come home, they would absorb roughly two-thirds of a sievert. That's significantly more than the average person receives while living on Earth (less than one-thousandth per year). Yet the National Cancer Institute has estimated an increased cancer risk by only three percentage points. Despite Kelly's experience, is it worth the risk?

Like Musk, Zubrin believes it is. "What it shows is that the cosmic ray dose on a Mars mission is not a showstopper," he said after the report was released.

Carberry thinks advances will be made to help shield humans from radiation but says that the bigger concern for space travelers is a system failure or getting hit by a meteorite. "When you're months and months away from Earth, it's those instantaneous things you have to be more worried about," he says. "Those are the real issues."

Maybe Carberry is right and radiation shielding will be vastly improved by the time humans can visit Mars. But if not, it's a risk people will have to weigh.

"If you're older, you shouldn't care," Pascal Lee says. "If you're young, yes— you'll have a greater chance of dying of cancer than the average population, and you might die five to ten years earlier than you would've normally. But hey, you've gone to Mars!"

PASCAL LEE'S FIVE THINGS THAT WILL KILL YOU ON MARS, IN CHRONOLOGICAL ORDER OF CAUSE OF DEATH (EXCLUDING THE LACK OF DRINKING WATER AND FOOD):

1. **Low atmospheric pressure.** It's so low that, unless you have a spacesuit that's totally sealed, you're dead within seconds. And you're not conscious for most of those seconds.

2. **Composition of the atmosphere.** You can't breathe carbon dioxide. If that was all that was available to breathe, you'd be a goner in minutes. The brain would run out of oxygen and you'd die of hypoxia.

3. **Temperature.** The average is -80 degrees Fahrenheit. You'd last a few hours before exposure would kill you.

4. **Dust**. It's toxic. Ingesting its perchlorates and peroxides would kill you within weeks.

5. **Radiation.** If you avoided the first four death traps but didn't have proper shielding, radiation would kill you over the course of months or, if not, a few years.

Let's assume, for the sake of us becoming a multiplanetary species, that science satisfactorily solves the physical problems with a trek through the cosmos. How about the mental ones? Will passengers regress to incessant childhood cries of "Are we there yet?" Though the lengthy trip required by SpaceX's BFS sounds daunting, it's not exactly new to the human experience. Traveling from New York to California by boat in the mid-1800s meant journeying around Cape Horn at the southern tip of South America. It took anywhere from four to seven months in cramped quarters with no Netflix to binge watch, not a single phone to stare at, and plenty of rough waters to endure. Plus Dramamine was still a hundred years away from being invented. In 1804, Lewis and Clark spent eighteen months traveling by canoe and foot from Illinois to the western edge of the United States. All in the name of exploration.

When you look at it that way, six months in a spacecraft designed with some degree of modern comforts shouldn't be too much to tolerate. Scientists have already learned to grow lettuce on the ISS, so the crew will be able to grow fresh vegetables and add a little variety to their freeze-dried rations. They can wash down their salads and stay hydrated with clean water processed from their own urine. That urine could even be mixed with nitrates to help their spacecraft gardens grow. It may not sound refreshing or delicious, but once they get over the thought of food and drink made from their own pee, there won't be any bad aftertaste. Plus, it's a waste management solution—where else would you put all that urine? The ISS has also helped prepare astronauts for spending way too much time together. But Mars is still a dreadfully long trip for a small group who may eventually tire of one another's stories, quirks, and various smells. Help might come from a special crew member who won't have any of those problems: an artificial intelligence robot.

It sounds like HAL from *2001: A Space Odyssey*, but the good news is that AI has advanced well beyond Stanley Kubrick and Arthur C. Clarke's futuristic, sinister vision of the technology. Not only can it offer predictive maintenance aboard the spacecraft, which will come in handy when you're flying to Mars and want to avoid getting stuck in the biggest middle-of-nowhere ever experienced, but it can even act as a companion and help resolve issues with crew members.

IBM has been working with the German Space Agency and Airbus to develop just such a robot using its Watson AI technology. They call it CIMON (Crew Interactive MObile CompanioN). The floating robot, shaped like the world's smartest basketball, took its first trip to space in 2018 aboard the ISS as an experiment in assisting with maintenance, analytics, and experiments—and, perhaps more importantly, understanding language, feelings, and emotions.

"The real idea behind CIMON was more on the isolation and stress reduction part," explains Matthias Biniok, Lead Watson Architect at IBM. "CIMON can monitor the health and stress of astronauts and act accordingly."

To do so, this super-complex system of data and algorithms communicates through smiles, winks, and speech from a face that looks like it might have been designed for a first-generation Nintendo Gameboy. Maybe that will make it more disarming as it listens to astronauts and discusses their feelings with them. In theory, it could help resolve or even prevent conflicts between crew members who might disagree on points regarding the mission or who simply need a break from each other. That sort of help can be useful once Earth is entirely out of view and every gaze outside is a constant visual reminder of how far from home they truly are. Such an experience could have a unique psychological effect. Humanity has coped with new experiences with every step forward, but this leap will be unlike any before.

Another way to help create a smoother journey is simply to give engineers more time to develop advanced propulsion systems, like ion thrusters and nuclear thermal engines. The latter could trim about six weeks off the journey. Less time means less of all the bad stuff: cosmic cancer rays, muscle atrophy, loss of bone mass, stress, and fewer chances of getting hit by giant space rocks. The faster we can get to Mars, the better.

"The psychological benefits of seeing green growing things up there is something we can almost not speak highly enough of, because when you're so far away from Earth on one of these very long duration flights, you're not going to see the planet for a long time. Having a garden, having that little piece of Earth with you, could be really important."

—**Dr. Gioia Massa,**
Life Sciences Project Scientist,
Kennedy Space Center, in 2019

ENTER
THE BRAHMANAUT

As the U.S. prepares to send people to Mars, officials will have to find candidates who are mentally and physically fit for the journey. The question is: Do they have to be American? Or, instead of sending astronauts, can we send Brahmanauts?

A presentation at the sixty-sixth International Astronautical Congress in 2015 proposed the idea of Brahmanauts—explorers trained in yoga, meditation, and ancient Vedic traditions—for deep space exploration missions. Conceived by Chrishma Singh-Derewa, a systems engineer at NASA JPL, in partnership with the Indian Space University and a group of researchers, the notion is based on the Brahmans' expertise in centuries-old practices that allow them to sit still for years, subsisting on minimal calories per day and decreased oxygen consumption. Both help lower the costs for a Mars mission substantially. Furthermore, yogic exercises not only tone the digestive system but also strengthen the abdominal muscles, ventilate the lungs, and increase the oxygenation of the blood. They also tone the nervous system, which would mitigate the physical stress on the body's fluids due to microgravity. All are ideal benefits for what could be a year-long round trip.

"A journey into no thought and the ability to clear and re-center one's mind is critical to surviving and operating under the extreme conditions," the paper states. "Brahmanauts have a spiritually strong center behind all they do. It is this inner purpose that enables them to overcome the physical and psychological barriers of long duration spaceflight, in addition to discovering the internal energies required to reach into the great unknown."

Staying sane and healthy is key, but Brahmanauts would also be trained in the space sciences and requirements necessary to complete the mission and return safely.

LIFE ON MARS?

Space travel has its challenges, but if things work out with the big SpaceX rockets or other types of spacecraft, humans could soon be ready to migrate to Mars. Maybe, if Elon Musk has his way, those humans could be our kids. As parents, that would mean one thing: we'll worry. What will they eat? Will they be warm enough? What will they live in? And would you be crazy to let them go? Dr. Bill Nye, aka "The Science Guy" and CEO of the Planetary Society, thinks you would, given that it's freezing, there's hardly any water, no food, and nothing to breathe.

"This whole idea of terraforming Mars, as respectful as I can be, are you guys high?" he asked an interviewer in 2018. "We can't even take care of this planet where we live, and we're perfectly suited for it, let alone another planet."

Others have a bit more hope, like Robert Zubrin, whose plan lays out detailed plans for terraforming. One method is through a system of orbiting mirrors. Imagine space-based reflectors—more than 150 miles in diameter—bouncing sunlight much closer to the Martian poles. "A four-degree Kelvin temperature rise imposed at the pole should be sufficient to cause the evaporation of the carbon dioxide reservoir in the south polar cap," Zubrin wrote. They'd be similar perhaps to what Kim Stanley Robinson imagined in *Red Mars*, but far more advanced than the scheme William Pickering and Charles Cros devised to reflect sunlight through space and signal Martians.

Producing greenhouse gases, Zubrin added, could help warm the planet within several decades, and although humans still couldn't breathe the air, they could at least trade their spacesuits for sweaters and gas masks to take a walk under the red sky. That'd be as good as it gets until Mars is oxygenated, which, the engineer theorized, could take nine hundred years. It's a generous estimate. Others believe terraforming could take a bit longer.

Dr. Chris McKay, senior scientist at NASA's Ames Research Center, believes warming the planet could take a century, but oxygenation could take a hundred thousand years or longer, "unless one postulates a technology breakthrough." These estimates depend on answers to key questions, including the amount of water, carbon dioxide, and nitrate in the soil. McKay called *Curiosity*'s discovery of the latter encouraging.

Carl Sagan explored the idea of terraforming in his 1980 book *Cosmos*, and though he didn't specify a hundred thousand years, he also believed the process could take thousands of years. If it were to work, he imagined that Percival Lowell's vision of Mars may finally come to pass. We, the new Martians, might need to transport water from the melting polar

> "THE FIRST ASTRONAUTS TO REACH MARS WILL PROVE THAT THE WORLDS OF THE HEAVENS ARE THE CASE FOR MARS *ACCESSIBLE* TO HUMAN LIFE. BUT IF WE CAN TERRAFORM MARS, IT WILL SHOW THAT THE WORLDS OF THE HEAVENS THEMSELVES ARE *SUBJECT TO* THE HUMAN INTELLIGENT WILL."
>
> **—Dr. Robert Zubrin,**
> *The Case for Mars,* 2011

caps to other regions of the planet—and we'd do that by building a network of canals.

Given the extremely long timeline for terraforming, Pascal Lee notes that it would require dedicated, focused groups of people carrying on the work throughout the centuries, with consistent government backing:

> *"Terraforming Mars would take the taxes of everyone on Earth for generations to make it happen. You might wonder, well what for in the end? To create yet another taxable planet? I don't believe in terraforming. I think it's physically possible, marginally plausible from an engineering standpoint, but a complete non-starter from a societal, or political standpoint. From that standpoint, I'd say it's impossible."*

Maybe Bill Nye is right.

Dr. Jim Logan, a former NASA flight surgeon, certainly thinks so. He believes, like Elon Musk and Stephen Hawking, that our species must become multiplanetary to survive. But not on the Red Planet. Through lectures across the world and via his company, Space Enterprise Institute, Logan combats hype with realism to remind people of how dangerous Mars would be. To think that humans could simply terraform the place or erect a big shield to protect us is "magical thinking"—a concept he defines as "ignorance to the power of arrogance."

"Mars is the most tragic planet in the solar system," Logan says. "It started off with an atmosphere. It started off being wet. It started off with flowing water. And in a slow-motion death spiral, it basically became nothing but a thin, cold, desiccated, radiation-infested dead planet. Mars is the most miserable place I could ever imagine to put human beings. Ever."

Rather than make it less miserable, we could likely progress faster in genetics and make ourselves better equipped to survive on Mars as it is. While we're at it, let's finally cure cancer and other diseases, too. If that's just more magical thinking, perhaps we should accept that Mars is best suited as an outpost for science. If we desperately need a new place to live we could search for one on another Earth-like planet with a lovely, ready-to-breathe atmosphere. We just need to master interstellar space travel first. And that will take some patience.

In the meantime, centuries of science-fiction stories have romanticized Mars and created such wonder that it's become easy to overlook the obvious problems called out by Nye, Lee, and Logan. We want to begin our own Martian chronicles and we want to be Mark Watney

"If you're trying to draw an analogy about people trying to start a new life in a place with a promise of abundance and safety and happiness, Mars is really not that. Mars is a God-forsaken place. It just looks like Arizona or Utah, but it's not Arizona or Utah. It's completely deadly."

———————————————

—Dr. Pascal Lee,
on the idea of settling Mars
versus exploring it, 2019

without being trapped alone. But adding to the long list of challenges are recent studies suggesting that Mars does not have enough carbon dioxide to be terraformed, unless scientists are able to drill deep into its interior and discover a reservoir of carbon just waiting to be released into the atmosphere. Until then, if your kids go to Mars, they're going to do a lot of whining.

Let's start with the places they might live. How will contractors build housing and any other kind of structure needed to begin a civilization? Just like on the ride over, Martian immigrants will need to be properly shielded so they don't get roasted by cosmic radiation. That could mean burrowing into the side of a hill or the rim of a crater. In fact, crater-view real estate could become the first interplanetary hot spot. One way or another, they'll need

to be living underground like earthworms (marsworms?) or moles.

If you ask space architect John Spencer, the way to build on the Red Planet is by employing swarms of small robots controlled by AI. Mechanical worker bees, so to speak. And worker ants. It sounds like sci fi, but the technology is already being studied and it's far more efficient than lugging standard Earth-based construction equipment all the way to Mars.

"In my mind, they'll be a key construction method used for space," Spencer says of the swarms. "Why would you take a bulldozer, schlep it to space, and soft land it, when in the future you could have a thousand small robots about the size of your hand with different tool attachments, whether it's welding, or digging, or carrying? And they'll work in swarms to

Rendering of an Operational Mars Base.

build and move and dig and tunnel. They can work twenty-four hours a day if they're charged and maintained by other robots."

European scientists began exploring these possibilities more than ten years ago with a project called I-SWARM. In addition to tiny automated machines, the swarm includes robots that could configure themselves into larger robots to take on more complex tasks. Like a Martian Voltron.

"We now know there is water and dust, so all they would need is some sort of glue to start building structures, such as homes for human scientists," Marc Szymanski, a robotics researcher at the University of Karlsruhe in Germany, said in 2008.

Imagine them all working harmoniously and industriously like real, nonrobotic bees and ants, but instead of building anthills and beehives, they're building whatever we program them to build while we sit back and watch and not soak in cancer rays. Such robots could get help making bricks from 3D printers using indigenous materials on Mars. Once materials are available, they can build arches or whatever types of designs fit your Martian crater mansion needs and personal tastes.

"You have to really think differently and cleverly for how you would actually develop for space," Spencer emphasizes.

That's the attitude taken by a design exploratory called Mars Ice House. It proposes 3D printing above-ground homes with, as the name suggests, ice. Water is a building block of life—so why not build houses as well? These Martian igloos would allow natural light to brighten the interior, plus the ice would serve as a radiation barrier. The plan also suggests a vertical hydroponic greenhouse to offer luxurious oxygen and food—ideal for new homeowners not wishing to die right after moving in.

Once craters are populated and ice houses stand triumphantly above the surface, how will the kitchens get stocked? Will they be packed with potatoes, like the ones grown by Mark Watney in *The Martian*? Since the soil is full of toxic chemical compounds, scientists would need to do a lot more than fertilize it with feces to make it viable. Growing food on Mars will be a critical part of any prolonged stay. Vegetables could be cultivated hydroponically, as they are on the ISS. Growth chambers

A rendering of a deployment and inflation of habitat modules. From right to left, the modules are inflated and connected by airlock.

An example of a Martian habitat 3D-printed with ice called Mars Ice House.

could be sent in advance, so healthy greens could be ready for humans by the time they arrive—along with an oxygen supply generated by the plants. As time and research march on, carnivores may even be able to enjoy some fresh Martian farm-to-table meals. And to wash it all down, groups in both California and the country of Georgia are already exploring ways to grow grapes for wine.

There you have it. If all the stars align, your kids could eventually be living like moles inside a crater or a block of ice and toasting to their new planet with a glass of Martian wine. Now that that's sorted out, when are you going to have Martian grandkids?

"VIRTUALLY ALL CURRENT PLANS FOR HUMAN ACTIVITIES ON MARS ASSUME THAT EARTH LIFE WILL GROW NORMALLY IN THE PARTIAL GRAVITY OF MARS. THERE IS ESSENTIALLY NO DATA TO SUPPORT THIS."

—**Chris McKay,**
Senior Scientist at NASA's
Ames Research Center, 2019

A TRIP TO MARS
WITHOUT THE DEADLY PARTS

To some, Las Vegas is about more than gambling and shotgun weddings. It's about visiting Paris, New York, Rome, Egypt, and Venice without walking more than a few blocks. Soon, tourists will even be able to stop in on Mars.

Mars World, currently under development, is a $2 billion domed entertainment park that will offer Earthlings the chance to experience the Red Planet without any of that pesky long-distance space travel, cosmic radiation, or cruel frigid temperatures.

The extraterrestrial experience is the brainchild of space architect John Spencer, who pictures visitors beginning their adventure by being transported to the year 2088 (by Earth's calendar). In that future world, they'll find themselves on the rim of a Martian crater before heading underground to immerse themselves in an extraterrestrial culture—all based on real science.

Guests will have a chance to experience Martian gravity, gaze at a different colored sky and many more stars, meet Martian characters, see exotic plant life, watch Martian sports and TV, and enjoy Martian food and drinks.

"Beer will be light because of low gravity," Spencer jokes. Developing Martian humor will be just another aspect of building an alien society.

Mars World will also help humans develop real Martian civilizations for whenever they may arise. "We want to build a Mars World community where our visitors can participate in design competitions, research, and really use the entertainment park to develop the real thing," Spencer says. "It's a positive view of the future that you can help make happen."

HOPPY PRICE'S MARTIAN MENU

Engineers devote a great deal of time to figuring out the logistics of interplanetary space travel, like how to make rocket fuel on Mars so that return trips to Earth are possible. "To me, a much more important problem to work, and a higher priority, is growing food on Mars," says Hoppy Price, chief engineer for the Mars Exploration Program, "because I think the psychological benefit to the crew is much greater."

Growing your own fresh food would provide a welcome break from packaged Earth foods, freeze-dried whatever, and other meals that are designed to last for a couple years before being eaten. Astronauts on the ISS have enjoyed a psychological boost just from growing hydroponic lettuce. But would it be just a vegetarian diet? Maybe not. As Price explains:

> "When you look at raising farm-type animals, you'd want to look for something that doesn't require a lot of resources. "Having cows grazing on Mars would be such a huge investment of resources. You'd have to have such a big area. Then you'd have cows living in 1/3g—how well would they do? It seems much more likely to raise fish on Mars. I think fish would do well, especially if you can mine the ice on Mars—the freshwater glaciers. You should be able to theoretically have plenty of water, so you could have fish farms and eat fish. Have caviar. Maybe raise shrimp. All those seem much more likely than other types of farm animals. Maybe you could do chickens and grow seeds for them to eat and have eggs. Chickens probably don't require as much of an area of land for them to thrive in. It'd be interesting to wake up to a rooster crowing on Mars."

THE JOY OF MARTIAN SEX . . . OR NOT

If humanity is going to colonize Mars, we'll need to be able reproduce there. The question is, can we do "it" in space? We don't know, because human sex in space has not been studied yet. What we do know is that trying to find out raises many important questions and ethical issues, once again thanks to humanity's nemeses: radiation and microgravity.

If microgravity has such a strong effect on the adult bodies of astronauts, what might it do to a developing baby? And before we get ahead ourselves, could a woman even get pregnant in such an environment? Mars doesn't have quite the zero-g issue that the ISS or interplanetary travel has, but its gravity is low—about one-third that of Earth (0.38 g). That's a major change for our species to adapt to.

All life on Earth, from single cells to everything living today, has evolved in a 1g environment. "Every single other thing about that environment has changed," says Dr. Alexander Layendecker, a sexologist with a professional background in space operations who has researched human reproduction and sexual health factors in outer-space environments. He notes that "the land masses have changed, the atmospheric composition has changed, the climate temperatures have changed, the weather patterns, the creatures themselves and the environments they're within, all of that has altered significantly over time—often very gradually, sometimes more rapidly, but gravity has always been the one constant. So to suddenly and radically change that constant could potentially have profound effects on our cellular development."

Not much has been done to find out what those effects would be, aside from some experiments with rats and mice. In 1979 Russian scientists brought rats aboard the Cosmos 1129 mission for a space sex study. The females ovulated and the males did their part, but no rats got pregnant.

More astrorats scurried their way to space in 1998, along with a few crickets, snails, and fish, for sixteen days aboard the space shuttle *Columbia*. But this time the rats were babies. Researchers wanted to find out if zero gravity would affect their neurological development. During that short journey, a group of eight-day-old rats would develop from infants to adolescents, meaning a lot of maturity would take place in their nervous systems. Many of them, however, did not survive the flight. That's a pretty clear indication that microgravity is bad for babies. A second group of fourteen-day-old rats lived, but showed difficulties with their cardiovascular systems, their muscular development, and the way their nerves connected to their brains. Microgravity is bad for toddlers too.

More than a decade later, in 2013, the Japanese Aerospace Exploration Agency (JAXA) partnered with NASA on its "Space Pup" experiment aboard the ISS to see if radiation would screw up the DNA of freeze-dried mouse sperm. After nine months, the space sperm returned to Earth and was injected into fresh eggs and transferred to surrogate mice moms. Amazingly, happy, healthy space pups were born. There were no apparent signs of abnormalities. So, that's good news.

Assuming a pregnancy can occur in lower gravity, and a baby is born shielded from radiation in a dusty crater-side mole hole, it's possible that deviations within its cellular development could start as early as the third trimester, when gravity helps infants orient properly in the womb. Once born, its bones would likely be elongated and have a lighter density. Maybe its brain would grow larger, too. All those tall, big-headed Martians that Hugo Gernsback and others imagined long ago would become a reality, to an extent, as human Martians. If they were ever to make a trip to Earth, they'd need some form of spacesuit to survive; an exoskeleton of sorts to help the blood flow properly and to protect them from all our nasty germs they would have never been exposed to—like the ones that killed off H. G. Wells's Martians.

Testing off-world reproduction is no easy feat. To do so would be to knowingly put lives at risk. We don't know how it might affect the mother, and it seems quite clear that exposing a baby to a high-radiation environment is really bad parenting. The child would undoubtedly endure a lifetime of physical and mental developmental effects. So, from an ethical standpoint, it's a tough mission to greenlight. Of course, a pregnancy could happen by accident. To date, that's been avoided, despite a few opportunities for some low-earth orbit hanky-panky, as when Mark Lee and Jan Davis became the first married astronaut couple to go to space together back in 1992. They were married in secret just before their launch date and NASA didn't have time to train a substitute. But they made no claim to fame upon their return home eight days later. Rumors have been floated that Russian cosmonauts may have done the deed. During Valery Polyakov's record stint in space, he was offered a sex doll to help him cope with the lengthy stay. He opted for a collection of porn instead. But when Elena Kondakova joined him on the Mir space station, some believed he may have put the porn aside. Both have denied indulging in any flings.

Though NASA doesn't officially forbid sex in space, astronauts have claimed to maintain strictly professional relationships. Finding answers might require the help of people in nonprofessional relationships, like space tourists. That's likely a few years off, but a curious couple excited about a spaceflight may wish to make love to make history. If it happens on Virgin Galactic, Branson may wish to mark the occasion with a name change.

In the meantime, add space sex to the long list of challenges to overcome. "It's like doing a 500-piece puzzle, but we only have three of the pieces right now," Layendecker says. "It's not really looking great."

A company called SpaceLife Origin is trying to add a few more pieces to solve the puzzle. By 2024 it plans to deliver the first baby in space, 250 miles above Earth. The mission will involve sending a particularly brave pregnant volunteer into the heavens accompanied by a medical team prepared to mitigate all risks. Her pregnancy will have been monitored to ensure that the baby was developing perfectly prior to blastoff.

So, what kind of mom would skip the hospital in favor of someplace not on Earth? It won't be just any woman looking to give her child an immortal place in history. According to Dr. Egbert Edelbroek, the company's chief strategy and innovation officer, several criteria will have to be met. The mother will have to have given birth twice already without any difficul-ties. This, Edelbroek says, reduces the risks of pregnancy complications by 95 percent. Although extra radiation shielding measures will be taken, SpaceLife Origin also plans to select couples with a high natural radiation resistance and a higher DNA repair mechanism to deal with any radiation received. They'll even test the cell tissues of the candidates' two children to ensure that the mix of genes resulted in a similarly high natural resistance. In short, they're looking for Super Mom. Ideally, the couples who make the final cut will have a mix of regions, races and religions. "We want this step in evolution to be owned by as many groups as possible," Edelbroek says.

While it'd be great to have a beautiful space station to give birth on, another existing option is the Sierra Nevada Corporation's

Dream Chaser. This small space shuttle can carry up to seven people and would reenter the atmosphere at 3g, which would be significantly lower than the usual 5g–6g reentry that trained astronauts experience. Though not ideal, Edelbroek believes it will work and that within the next few years the g-forces will improve as space tourism continues to develop. "They will want to transport rich people who can afford it but are not willing to get a two-year training and get in healthy shape," he says.

If Edelbroek is right and delivery is successful, Space Baby will be flown back to Earth and demonstrate that an off-world birth is possible and safe. The next focus would be exploring the possibilities of pregnancies in the cosmos. SpaceLife Origin will also be testing embryo development in space to learn how much gravity is required. Can it develop under the Red Planet's lower gravity level? If not, humanity will have to continue exploring ways to multiply in the universe. Like, inside the moon. And not our moon—Deimos, one of the two moons of Mars.

"IT'S A SMALL STEP FOR A BABY, BUT A GIANT BABY-STEP FOR MANKIND."

—Dr. Egbert Edelbroek,
SpaceLife Origins
Chief Strategy & Innovation Officer, 2018

SPACE BABY, THE EXTRATERRESTRIAL

Once a baby is born in space and returns to Earth, what nationality will he or she be? As of now, it remains unexplored legal territory. "Formally it would still be an alien," Edelbroek suggests.

GOODNIGHT, MOON. GOODNIGHT, PEOPLE LIVING INSIDE THE MOON.

As John Spencer said, the challenges of space require clever thinking. So, here's an idea: hollow out Deimos and fill it with rotating super space stations that create artificial gravity. It's a plan proposed by Jim Logan and Astrodynamics Consultant Daniel Adamo as a way to create a viable place to live very close to Mars. The surrounding moon would provide all the radiation protection that humans would need. Just like that, you'd have a safe home with a simulated Earth environment. That means horny colonists could try to make all the space babies they want and give them a secure place to grow.

The space station idea actually dates back to the early 1900s when Russian physicist Konstantin Tsiolkovsky explored off-world living in his scientific fantasy novel, *Beyond the Planet Earth*. In the story, overpopulation forces people to look for new places to live. Rather than settle the moon, Tsiolkovsky proposes the idea of colonies on asteroids with human-made gravity to replicate an Earth-like environment. "Nothing could be simpler than to create it artificially, you see, by rotating the house," he wrote. "In space, once you start a body rotating, it goes on rotating indefinitely, there is no effort involved; so the gravity is also maintained indefinitely, it costs nothing."

In the mid-1970s, Princeton physicist Gerard K. O'Neill proposed space colonies as well, but not as fiction. Instead, he believed that creating such a living space was perfectly doable with existing technology and felt confident it would happen before 2005. Like Jeff Bezos—who, it's worth noting, studied under O'Neill at Princeton—O'Neill had concerns about solving the energy crisis on Earth and proposed the colonies as a viable solution. And like Tsiolkovsky's story, O'Neill's plan would also help alleviate the planet's rapidly expanding population.

"On the inside of a hollow, rotating vessel the gravity can be made to be the same as on Earth, and if the vessel is big enough, the human body will find the artificial gravity indistinguishable from the real thing," O'Neill suggested.

In one of his designs, he imagined a sphere with a diameter of more than 1,500 feet that could hold up to 10,000 colonists. To put that size in perspective, the Empire State Building stands 1,454 feet at its tip. The gravity created by the sphere's centrifugal force would keep everyone's feet on the ground so they could enjoy comfortable lives in large private

Rendering of a pair of O'Neill cylinders

apartments with sunlit gardens, a shallow river flowing near the equator, with beaches and recreational paths for bikers and joggers. An agricultural area would grow food to help the colony be as self-sufficient as possible. To anyone stuck living in a Martian crater it'd be like a cosmic version of Beverly Hills.

Logan estimates that eleven of these swanky O'Neill colonies could eventually fit back-to-back inside a single coring of Deimos. Sunlight could be piped in through solar tubes and nuclear energy could help produce light and power. But of course, before any space colonies can be squeezed inside the moon, there's that whole issue of hollowing it out. That alone sounds like a daunting task, but when confronted with that concern, Logan has a simple response: "Any intelligent civilization that can travel 560 million kilometers to a destination and is unwilling or unable to dig a hole doesn't deserve to be there in the first place."

All that radiation protection the colonists would enjoy is great, but they'll need it on the way there, too. Sure, NASA and SpaceX are working on this knotty problem, but Logan and Adamo have started a plan of their own. It's a spacecraft, spherical in shape, that's designed with an innovative concept: water. By using liquid water as a propellant for a nuclear thermal propulsion system, they can double the efficiency achievable with chemical propellants. Plus, the water serves as a beverage and surrounds the crew compartment like a jacket to protect them from the radiation. They call the spacecraft *Aquarius*.

So, how long would it take to settle Deimos? "Twenty years after the laughing stops," Logan says, pointing out that history has proven that equation to be true over and over. For example,

the idea of controlled flight in a heavier-than-air machine was crazy talk until the 1880s, when a few scientists stopped giggling about it and started applying mathematics to the problem. By 1903, the Wright Brothers achieved flight for the first time. Less than twenty-five years later, Charles Lindbergh flew across the Atlantic. Forty-two years after that, Neil Armstrong reached the moon. When NASA, private companies, or other countries decide that Logan's right and set their sights on Deimos, they'll eventually get there and it'll become a celestial stepping stone to Mars.

"The history of the human species won't really begin until we leave the planet in numbers sufficient to build self-sustaining and

Color-enhanced views of Deimos, taken on Feb. 21, 2009, by the High Resolution Imaging Science Experiment (HiRISE) camera on NASA's Mars Reconnaissance Orbiter.

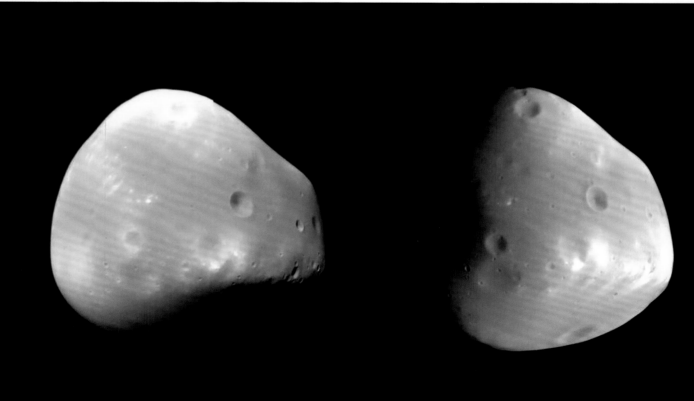

self-replicating human communities in space. Period," Logan says. "Otherwise we'll just be a historical footnote in galactic history. Single planet species don't survive."

If the laughing stops now, we'll be well ahead of Stephen Hawking's timeline.

ANOTHER GIANT LEAP

Konstantin Tsiolkovsky once said, "The Earth is the cradle of humanity, but mankind cannot stay in the cradle forever." We humans have a few problems to solve before leaving our planet, but even if a lot of the current hype might be nothing more than just that, it serves us all well as fuel for finding the solutions we need. It's what we've always done, going back to whoever first figured out how to make a boat and crossed that great unknown we call water. When we do it again, the first man or woman to set foot on Mars, or even inside Deimos, will be etched into history as unforgettably as Christopher Columbus and Neil Armstrong.

One other thing we can be sure of is that wherever humans colonize away from the Earth, they need to act much as the Europeans did when settling the New World. They must make use of natural resources. Europeans didn't bring planks of wood, buckets of nails, and other building materials, and they certainly didn't pack loads of lunches. Future settlers won't be able to do that either, but they will be armed with extensive knowledge and as many tools as possible, ready to live off whatever land is there. It would be the culmination of Earth's obsession with Mars over the millennia, from our ancient ancestors who spied a glimmering red dot in the night sky to Percival Lowell's insistence on canal-digging Martians to David Todd's desperate attempts to catch their messages from a hot-air balloon to the robots roaming the surface right now.

As we traipse across the alien landscape and mark our territory, we'll be able to study the Red Planet up close and personal in hopes of finally discovering the elusive answers to the origin of life in the universe. Maybe we'll even learn that it all began right there on Mars, when it truly was different place, ruled by happy microorganisms without a care in the world. What if a few of those little Martians hopped aboard a meteorite to Earth and over billions of years prospered into life as we know it? All that time we humans have spent wondering, stargazing, creating characters and mythologies, and meticulously calculating interplanetary flight paths would have achieved a purpose we never expected: finding our way home.

MARTIAN RESOURCES

Mars has enjoyed a lot of coverage ever since humans discovered the written word. In addition to numerous interviews with members of the science community, the following articles, books, websites, and videos were valuable resources in my research for writing this book.

BIG HOPES AND BIGGER MISUNDERSTANDINGS

"Along a Ray of Light." *Anaconda Standard*, May 23, 1897.

Associated Press. "Mars Has Animals, Professor Infers." *Evening Star*, October 28, 1926.

———. "25-Mile Diameter Beacon Required for Mars to Flash Earth." *Evening Star*, July 23, 1939.

"Astronomers Now Talk of Signalling Planet Mars." *New York Journal and Advertiser*, January 1, 1901.

Bakich, Michael E. *The Cambridge Planetary Handbook*. Cambridge University Press, 2000.

"Blue light signals from Mars." *Evening Star*, July 23, 1939.

"Canals on Mars are all Nonsense." *Washington Herald*, February 27, 1916.

Carrington, Hereward. *Eusapia Palladino and Her Phenomena*. New York: R. W. Dodge, 1909.

Crowe, Michael J., editor. *The Extraterrestrial Debate: Antiquity to 1915*. University of Notre Dame, 2008.

Dick, Thomas. *Celestial Scenery; or, the Wonders of the Planetary System Displayed*. Brookfield: E. and L. Merriam, 1838.

Elway, Thomas. "Do Beavers Rules on Mars?" *Popular Science Monthly*, May 1930.

Fisher, David E., and Marshall Jon Fisher. *Strangers in the Night: A Brief History of the Life on Other Worlds*. Washington, DC: Counterpoint, 1998.

Flammarion, Camille. *Dreams of an Astronomer*. New York: D. Appleton and Company, 1923.

———. "Mars and the Earth: Flammarion Compares the Conditions on the Two." *Chicago Daily Tribune*, June 14, 1896.

"From Schenectady, not Mars." *Literary Digest*, August 5, 1922

Haeusler, Dr. Rudolf. "Natural Science Gossip—Signals from Mars." *Waikato Argus*, August 5, 1901.

"'Hello, Earth! Hello!' Marconi believes he is receiving signals from the planets." *Tomahawk*, March 18, 1920.

"How to Signal to Mars: Wireless the Only Way Now, Says Nikola Tesla—Mirror Plan Not Practicable." *New York Times*, May 23, 1909.

Huygens, Christiaan. *Cosmotheros: Or, Conjectures Concerning the Planetary Worlds, and Their Inhabitants*. Glasgow, 1757.

"Interplanetary Radio Signals?" *Wireless Age* 7, no. 6, March 1920.

Ley, Willy. *Rockets, Missiles, and Space Travel*. New York: The Viking Press, 1957.

"Life on Planet Mars." *Minneapolis Journal*, December 14, 1901.

Lowell, Percival. *Mars as the Abode of Life*. New York: The Macmillan Company, 1908.

"Marconi Sure Mars Flashes Messages." *New York Times*, September 2, 1921.

"The Martians." "Pearson's Weekly" from *Lyttelton Times*, July 2, 1906.

"Martians Pelting Us: Wiggins Says Meteors Are Their Pictures." *New York Journal*, November 20, 1898.

"Martians Probably Superior to Us." *New York Times*, November 10, 1907.

"Martian Telegraphy Discredited." *New York Tribune*, January 12, 1901.

"My 34 Nights on Mars: How Prof. Edward S. Morse has been Studying the Great Planet through the Lowell Observatory Telescope and his own Interesting Account of What he Discovered there." *World Magazine*, October 7, 1906.

"The Name of God on Mars." *Kansas City Journal*, June 2, 1895.

Norton, Frank H. "How to Signal Mars." *Illustrated American*, June 9, 1894.

"One More Doubt." *Morning Journal Courier*, January 27, 1908.

"Queries for the Martians." *Cincinnati Enquirer* [from *Harper's Weekly*], May 18, 1909.

Sagan, Carl. *Cosmos*. New York: Ballantine Books, 2013.

"The Scheme to Signal Mars." *Popular Mechanics*, July 1909.

"Sending of Messages to Planets Predicted by Dr. Tesla on Birthday." *New York Times*, July 11, 1937.

Serviss, Professor Garrett P. "Professor Lowell's Last Conclusions About Life on Mars." *Los Angeles Examiner*, December 17, 1916.

Sincell, Mark. "Mars Flashes Earth Again." *Science*, June 12, 2001. https://www.sciencemag.org/news/2001/06/mars-flashes-earth-again (accessed June 1, 2019).

"A Society for Communication with Mars," *Literary Digest*, September 16, 1899.

Sullivan, Walter. "Nobel Prize Winner Believes that Mars Could Support Life." *New York Times*, August 8, 1965.

"The Talk of the Day." *New York Tribune*, May 30, 1901.

"Talk to Mars at $10,000,000 a Chat." *Columbian*, May 13, 1909.

Tesla, Nikola. "Talking with the Planets." *Collier's Weekly*, February 9, 1901.

"Venus Waiting for Her Chance." *Bridgeport Times*, April 22, 1922.

"What's the Matter with Mars?" *Washington Times*, April 17, 1904.

"Wiggins on the Aerolite." *New York Times*,

November 18, 1897.

"Would Signal Mars." *Chickasha Daily Express*, December 24, 1904.

SATAN, THE SPACE RACE, AND MARTIAN ROBOTS

ArtAlienTV. "The Martian - Humanoid Skull - Homo Aeolis." YouTube.com. https://www.youtube.com/watch?v=mzO6YsD2SjY&t=531s (accessed October 13, 2018).

Associated Press. "Life on Mars? Study hopeful, but scientists are snorting." *Cincinnati Post*, August 7, 1996.

David, Leonard. "Life on Mars? 40 Years Later, Viking Lander Scientist Still Says 'Yes'." Space.com, August 31, 2018. https://www.space.com/41689-nasa-viking-mars-life-search-gil-levin.html (accessed December 20, 2018).

"The Dead Planet." *New York Times*, July 30, 1965.

"Girl with Dreams Names Mars Rovers 'Spirit' and 'Opportunity'." NASA Press Release, June 8, 2003. https://www.nasa.gov/missions/highlights/mars_rover_names.html (accessed January 16, 2019).

Kluger, Jeffrey. "Here's What Explains the 'Morse Code' on Mars." *Time*, July 12, 2016. http://time.com/4402622/heres-what-explains-the-morse-code-on-mars (accessed March 3, 2019).

Lyons, Richard D. "NASA Hopes to Put Men on Mars in '80's." *New York Times*, May 27, 1969.

"Meteorite Yields Evidence of Primitive Life on Early Mars." NASA Press Release, August 7, 1996. https://www2.jpl.nasa.gov/snc/nasa1.html (accessed December 10, 2018).

Naeye, Robert. "Blazing a Trail to the Red Planet." *Astronomy*, October 1997.

Parsons, John. "Analysis by a Master of the Temple of the Critical Nodes in the Experience of his Material Vehicle." Circa 1949. https://hermetic.com/parsons/analysis-by-a-master-of-the-temple (accessed March 13, 2019).

Pendle, George. *Strange Angel: The Otherworldly Life of Rocket Scientist John Whiteside Parsons*. New York: Harcourt Books, 2005.

"President Clinton Statement Regarding Mars Meteorite Discovery." White House Office of the Press Secretary, August 7, 1996. https://www2.jpl.nasa.gov/snc/clinton.html (accessed December 10, 2018).

Sagan, Carl. "Going Beyond Viking 1: Touring Mars on Wheels." *New York Times*, August 11, 1976.

Sagan, Carl, and David Wallace. "A Search for Life on Earth at 100 Meter Resolution." Cornell University Laboratory for Planetary Studies, June 1970.

Simons, Howard. "Pictures Show Mars More Moonlike Than Earthlike, Fail to Disprove Life." *Washington Post*, July 30, 1965.

Squyres, Steve. *Roving Mars*. New York: Hyperion, 2005.

Straat, Patricia Ann. *To Mars with Love*. Charleston: Palmetto Publishing Group, 2018.

"Tributes to Terrorism Victims Are on Mars." NASA Press Release, September 8, 2011. https://mars.nasa.gov/mer/newsroom/pressreleases/20110908a.html (accessed January 7, 2019).

Vincent, Alice. "Sex, rocket science and Scientology: meet the other hidden figures behind Nasa's occultist foundations." *Telegraph*, February 17, 2017, https://www.telegraph.co.uk/films/2017/02/17/sex-rocket-science-scientology-meet-hidden-figures-behind-nasas (accessed on Oct. 11, 2018).

Wall, Mike. "A Schoolkid Will Name NASA's Next Mars Rover." Space.com, September 25, 2018. https://www.space.com/41920-nasa-contest-name-2020-mars-rover.html (accessed November 18, 2018).

Wilford, John Noble. "Viking Robot Sets Down Safely on Mars and Sends Back Pictures of Rocky Plain." *New York Times*, July 21, 1976.

CLOSE ENCOUNTERS OF THE MARTIAN KIND

Allen, Gracie. "'Flying Saucers' Long Stay Proves No Men from Mars Are Aboard." *Louisville* (KY) *Times*, July 14, 1947.

Allingham, Cedric. *Flying Saucer from Mars*. New York: British Book Centre, 1955.

Associated Press. "Flying Saucers Over Cincinnati." *Cincinnati Enquirer*, July 7, 1947.

———. "Stories Rival Flying Discs! Saucers 'Size of a House' Also Resemble 'Pie Pans.'" *Cincinnati Enquirer*, July 8, 1947.

Bowman, Tom. "Air Force says 'space aliens' at Roswell were test dummies 'Case Closed' report is unlikely to satisfy believers in UFOs." *Baltimore Sun*, June 25, 1997.

Carpenter, Major Donald G., editor, and Lt. Colonel Edward R. Therkelson, coeditor. "Introductory Space Science - Volume II, Chapter XXXIII, Unidentified Flying Objects, Department of Physics - USAF." Computer UFO Network. http://www.cufon.org/cufon/afu.htm (accessed November 2, 2018).

"Communicating with Mars." *Oamaru Mail*, July 5, 1909.

Edson, Arthur. "'Flying Saucers' Remain Unexplained Phenomena." *Hartford Courant*, December 15, 1957.

Flournoy, Theodore. *From India to the Planet Mars*. New York: Harper & Brothers, 1901.

———. *Spiritism and Psychology*. New York: Harper & Brothers, 1911.

"Flying Disk Gives Army Dizzy Whirl!" *Chicago Daily Tribune*, July 9, 1947.

"'Flying Saucers' Called Jet Planes by Officer— Still Mystery to Others." *Austin Statesman*, June 28, 1947.

Gaston, Henry A. *Mars Revealed; Or, Seven Days in the Spirit World: Containing an Account of the Spirit's Trip to Mars, and His Return to Earth; What He Saw and Heard on Mars, Etc*. San Francisco: A. L. Bancroft & Co., 1880.

Haines, Gerald K. "CIA's Role in the Study of UFOs, 1947–90." Central Intelligence Agency. https://www.cia.gov/library/center-for-the-study-of-intelligence/csi-publications/csi-studies/studies/97unclass/ufo.html (accessed September 30, 2018).

"Hairy Man from Mars Made but Brief Call." *Washington Times*, August 17, 1906.

Hartzman, Marc. "The London Lawyer Who Tried to Contact Mars via Telegraph." *Mental Floss*, February 7, 2018. http://mentalfloss.com/article/527779/london-lawyer-who-tried-contact-mars-telegraph (accessed February 24, 2020).

———. "Earth to Mars in the 1920s: The Strange Case of the Man Who Tried to Contact Martians via Radio." WeirdHistorian.com, February 19, 2018. http://www.weirdhistorian.com/the-strange-case-of-the-man-who-tried-to-contact-martians-via-radio (accessed February 24, 2020).

Heard, Gerald. *Is Another World Watching? The Riddle of the Flying Saucer*. New York: Bantam Books, 1953.

Hyslop, James. "The Smead Case." *Annals of Psychical Science*, August 1906.

"Makes Trip to Mars, See Race of Cyclops." *Salt Lake Tribune*, August 8, 1906.

McLaughlin, Commander Robert B. "How Scientists Tracked a Flying Saucer." *True*, March 1950.

Mike Wallace Interview, March 8, 1958.

https://www.youtube.com/watch?v=EZ-TMS81iSF4 (accessed November 1, 2018).

"A 'Nutty' Story? Unusual Objects Seen Flying in America." *South China Morning Post*, June 27, 1947.

Price, Harry. *Confessions of a Ghost-Hunter*. London: Putnam & Company, 1936.

"Radio Fans Duped by 'Men of Mars' in Flying Saucers." *Stars and Stripes*, July 16, 1947.

"Radio Yarn of Mars Men Riding Flying Saucer Brings Flood of Calls." *Hartford Courant*, July 15, 1947.

"Rancher Credited with Finding First 'Saucer' Now Sorry He Mentioned It." *Austin Statesman*, July 9, 1947.

Rosenberg, Daniel. "Speaking Martian." *Cabinet*, Winter 2000–2001. http://cabinetmagazine.org/issues/1/i_martian.php (accessed September 25, 2018).

"Saucer Pilots Could Be Smart Bugs or Plants." *Los Angeles Times*, March 14, 1950.

"Saucers? Maybe a Mighty Russian Throwing a Discus, Gromyko Hints." *New York Times*, July 10, 1947.

Swedenborg, Emanuel. *Earths in Our Solar System Which Are Called Planets and Earths in the Starry Heaven Their Inhabitants, and the Spirits and Angels There*. London: Swedenborg Society, 1962.

"Those Flying Saucers." *Globe and Mail*, July 8, 1947.

"Two Killed in Crash on 'Saucer' Mission." *Los Angeles Times*, August 3, 1947.

United Press International. "Lights Set to Attract 'Saucers.'" *Desert Sun*, October 20, 1973.

"Visits Planet of Mars; Finds One-Eyed Giants." *Idaho Recorder*, September 20, 1906.

MARS INVADES POP CULTURE

Alvear, Cecilia. "Martians Land in Quito." *HuffPost*, March 16, 2009. https://www.huffpost.com/entry/martians-land-in-quito_b_166776 (accessed February 24, 2020).

Associated Press. "'Mars Invasion' Heart Attack Fatal to Baltimore Man." *Washington Post*, November 13, 1938.

———. "Radio Play Terrifies Nation." *Daily Boston Globe*, October 31, 1938.

"BBC Horizon (1964) with Arthur C. Clarke Part 1 of 2." *Knowledge Explosion*, season 1, episode 6, September 21, 1964.

https://www.youtube.com/watch?v=K-T_8-pjuctM (accessed February 2, 2019).

Bradbury, Ray. *The Martian Chronicles*. New York: William Morrow, 2011.

Brown, Fredric. *Martians, Go Home*. New York: E. P. Dutton & Company, 1955.

Chandler, A. Bertram. *The Alternate Martians*. New York: Ace Books, 1965.

"Chris Hadfield on how David Bowie responded to Space Oddity cover and what it takes to be an astronaut." ABC News, August 22, 2017. https://www.abc.net.au/news/2017-08-22/chris-hadfield-david-bowie-and-how-to-become-an-astronaut/8830624 (accessed Jan. 6, 2019).

Clarke, Arthur C. *Sands of Mars*. New York: Pocket Books, 1954.

Crossley, Robert. *Imagining Mars: A Literary History*. Middletown: Wesleyan University Press, 2011.

Du Maurier, George. *The Martian*. New York: Harper & Brothers Publishers, 1897.

Gernsback, Hugo. "Do the People on Mars Look Like This?" *Nashville Tennessean and the Nashville American*, November 23, 1919.

———. "Evolution on Mars." *Science and Invention*, August 1924.

Greene, Richard. "When the Martians 'Invaded' Ecuador." *Argus*, May 14, 1949.

Greg, Percy. *Across the Zodiac*. Project Gutenberg eBook. November 21, 2003. http://www.gutenberg.org/cache/epub/10165/pg10165-images.html.

Heinlein, Robert A. *Stranger in a Strange Land*. New York: Ace, 2018.

Higgins, Chris. "Arthur C. Clarke Predicts the Future in 1964." *Mental Floss, June 15, 2015. http://mentalfloss.com/article/57157/arthur-c-clarke-predicts-future-1964 (accessed January 19, 2019).*

"Jeweler Dies from Stroke Suffered After Radio Play." Sun, November 13, 1938.

"'Mars' Radio Case Closed." *New York Times*, December 6, 1938.

Nicholls, Richard E. "The Biggest, Fattest Sacred Cows." *New York Times*, December 9, 1990.

O'Neil, Paul. "The Amazing Hugo Gernsback, Prophet of Science, Barnum of the Space Age." *Life*, July 26, 1963.

Philmus, Robert M., and David Y. Hughes. *H. G. Wells: Early Writings in Science and Science Fiction*. Los Angeles: University of California Press, 1975.

"Reactions to the Radio Panic." *Washington Post*, November 3, 1938.

"Rewriting Outrages H. G. Wells but Radio Scare Booms Book." *New York Herald*

Tribune, November 2, 1938.

Robinson, Kim Stanley. *Red Mars*. New York: Del Ray, 2017.

Serviss, Garrett P. *Edison's Conquest of Mars*. Los Angeles: Carcosa House, 1947.

Sokol, Tony. "Exploring David Bowie's Sci-Fi Fascination." *Den of Geek*, January 8, 2019. https://www.denofgeek.com/us/culture/david-bowie/253909/exploring-david-bowies-sci-fi-fascination (accessed January 10, 2019).

"Terror by Radio." *New York Times*, November 1, 1938.

Tevis, Walter. *The Man Who Fell to Earth*. New York: Del Ray Impact, 1999.

Two Women of the West. *Unveiling the Parallel: A Romance*. Boston: Arena Publishing Company, 1893.

"War of the Worlds." *Radiolab* podcast, March 24, 2008.

Weir, Andy. *The Martian*. New York: Crown Publishers, 2014.

West, Bruce. "Broadway Cashes in on Famed Radio Panic." *Globe and Mail*, November 11, 1938.

Wyss, Jim. "In Ecuador, after the green men from Mars invaded, the real tragedy began." *Miami Herald*, Feb. 21, 2014.

LEAVING THE CRADLE

Aldrin, Buzz. *Mission to Mars: My Vision for Space Exploration*. Washington, DC: The National Geographic Society, 2013.

Buckey, Jay C. Jr., MD, and Jerry L. Homick, PhD, editors. *The Neurolab Spacelab Mission: Neuroscience Research in Space: Results from the STS-90 Neurolab Spacelab Mission*. Houston: NASA, 2014.

Cabbage, Michael. "Lust in space: Study tells all." *Chicago Tribune*, March 11, 2001.

Chang, Kenneth. "Data Point to Radiation Risk for Travelers to Mars." *New York Times*, May 30, 2013. https://www.nytimes.com/2013/05/31/science/space/data-show-higher-cancer-risk-for-mars-astronauts.html (accessed November 12, 2018).

Crane, Leah. "Terraforming Mars might be impossible due to a lack of carbon dioxide." *New Scientist*, July 30, 2018. https://www.newscientist.com/article/2175414-terraforming-mars-might-be-impossible-due-to-a-lack-of-carbon-dioxide (accessed March 3, 2019).

Cucinotta, Francis A., and Eliedonna Cacao. "Non-Targeted Effects Models Predict Significantly Higher Mars Mission Cancer

Risk than Targeted Effects Models." *Nature Research*, article number 1832, May 12, 2017. https://www.nature.com/articles/s41598-017-02087-3 (accessed December 9, 2018).

Döpfner, Mathias. "Jeff Bezos reveals what it's like to build an empire and become the richest man in the world—and why he's willing to spend $1 billion a year to fund the most important mission of his life." *Business Insider*, April 28, 2018. https://www.businessinsider.com/jeff-bezos-interview-axel-springer-ceo-amazon-trump-blue-origin-family-regulation-washington-post-2018-4 (accessed December 3, 2018).

Hall, Loretta. "Setting the Record: Fourteen Months Aboard Mir Was Dream Mission for Polyakov." *RocketSTEM*, February 9, 2015. https://www.rocketstem.org/2015/02/09/russian-cosmonaut-valeri-polyakov-spent-record-breaking-14-months-aboard-mir-space-station-in-1990s (accessed May 27, 2019).

Hartmans, Avery. "The fabulous life of Amazon CEO Jeff Bezos, the second-richest person in the world." *Business Insider*, May 15, 2017. https://www.businessinsider.com/amazon-founder-ceo-jeff-bezos-early-life-2017-5 (accessed December 3, 2018).

Hawking, Stephen. *Brief Answers to the Big Questions*. New York: Bantam Books, 2018.

"James Logan, MD | Living on Mars." University of Michigan Engineering, January 26, 2016. https://www.youtube.com/watch?v=pPVORuanf18 (accessed December 10, 2018).

Kelly, Scott. *Endurance: My Year in Space, A Lifetime of Discovery*. New York: Vintage, 2018.

Kolbert, Elizabeth. "Project Exodus: What's behind the dream of colonizing Mars?" *New Yorker*, June 1, 2015.

Kornei, Katherine. "Sperm frozen in space produce healthy mouse pups." *Science*, May 22, 2017. https://www.sciencemag.org/news/2017/05/sperm-frozen-space-produce-healthy-mouse-pups (accessed January 7, 2019).

Lackey, Katherine. "Bill Nye: We are not going to live on Mars, let alone turn it into Earth." *USA Today*, November 19, 2018.

Layendecker, Alexander Benjamin M.P.H., D.H.S., A.C.S. "Sex in Outer Space and the Advent of Astrosexology: A Philosophical Inquiry into the Implications of Human Sexuality and Reproductive Development Factors in Seeding Humanity's Future throughout the Cosmos and the Argument for an Astrosexological Research Institute." Institute for Advanced Study of Human Sexuality, 2016.

Levy, Steven. "Jeff Bezos Wants Us All to Leave Earth—For Good." *Wired*, October 15, 2018. https://www.wired.com/story/jeff-bezos-blue-origin (accessed December 3, 2018).

"The Mars 100: Mars One Announces Round Three Astronaut Candidates." Mars One, February 16, 2015. https://www.mars-one.com/news/press-releases/the-mars-100-mars-one-announces-round-three-astronaut-candidates (accessed May 27, 2019).

McCandless Farmer, Brit. "Looking for life on Mars—at the bottom of a gold mine." CBS News, November 11, 2018. https://www.cbsnews.com/news/looking-for-life-on-mars-at-the-bottom-of-a-south-africa-gold-mine (accessed March 1, 2019).

McKay, Chris. "Prerequisites to Human Activity on Mars: scientific and ethical aspects." *Theology and Science*, May 2019.

McKay, Chris. "The Terraforming Timeline." NASA, 2016.

Mosher, Dave. "This guy invented a genius solution for pooping in space—here's how it works." *Business Insider*, February 16, 2017. https://www.businessinsider.com/nasa-pooping-space-challenge-winner-thatcher-cardon-2017-2 (accessed January 26, 2019).

O'Neill, Gerard K. *The High Frontier: Human Colonies in Space*. New York: William Morrow and Company, Inc., 1977.

Paoletta, Rae. "Elon Musk Still Isn't Answering the Most Important Questions About Mars. There's some life or death stuff to sort out." *Inverse*, September 29, 2017. https://www.inverse.com/article/36979-elon-musk-mars-colony-radiation (accessed December 6, 2018).

Rein, Richard K. "Princeton Connection Helps Amazon's Jeff Bezos Reach the Sky." Princetoninfo.com, December 5, 2018. https://princetoninfo.com/princeton-connection-helps-amazons-jeff-bezos-reach-the-sky (accessed February 10, 2019).

"Robotic Ants Building Homes on Mars?" *Science Daily*, October 27, 2008. https://www.sciencedaily.com/releases/2008/10/081021190644.htm (accessed April 4, 2019).

"Space Poop Challenge." NASA, October 12, 2016. https://www.nasa.gov/feature/space-poop-challenge (accessed January 25, 2018).

Singh-Derewa, Chrishma, Dr. Jugvir Singh, Poonampreet Kaur Josan, Srikanth Raviprasad, Priyanka Srivastava, Rahul Goel, and Maitreyee Sharma Priyadarshini. "An Initiative to Nurture 'Brahmanauts' For Future Human Space Flight." International Astronautical Federation, 2015.

Zubrin, Robert, with Richard Wagner. *The Case for Mars: The Plan to Settle the Red Planet and Why We Must*. New York: Free Press, 2011.

INDEX

PHOTO CREDITS

tian Odyssey and Others (Fantasy Press, 1949), numbered, limited edition. From the collection of Barry Abrahams **TOP RIGHT** Isaac Asimov, *The Martian Way and Other Stories* (Doubleday, 1955), first edition. From the collection of Barry Abrahams **BOTTOM LEFT** Donald Wollheim, *Secret of the Martian Moons* (John C. Winston, 1955) first edition. From the collection of Barry Abrahams **BOTTOM RIGHT** Leigh Brackett, *The Sword of Rhiannon* (Boardman, 1955), first British and first hardcover edition. From the collection of Barry Abrahams

172 Poster for *The Wizard of Mars*, 1965. Everett Collection

173 Still from *Devil Girl from Mars*, starring Patricia Laffan, 1954. Everett Collection

175 Poster for *Frankenstein Meets the Space Monster*, 1965. Everett Collection

177 Marvin the Martian, undated. @ Warner Bros. / courtesy Everett Collection

178 Still from *Abbott and Costello Go to Mars*, 1953. Everett Collection

179 Still from *The Three Stooges in Orbit*, 1962. Everett Collection

180 U.S. poster for *Santa Claus Conquers the Martians* **TOP LEFT:** John Call, Victor Stiles, Donna Conforti, **top right:** Bill McCutcheon, Leila Martin, Pia Zadora, 1964. Everett Collection

183 Fredric Brown, *Martians, Go Home* (E. P. Dutton & Company, 1955), first edition. Author's collection

184 *My Favorite Martian*, starring Bill Bixby (left) and Ray Walston (right), 1963–66, first episode. Everett Collection

185 *Destination Earth*, Sutherland Productions, 1956.

187 *The Twilight Zone*, "Mr. Dingle, the Strong," Season 2, Episode 19, March 3, 1961. Photofest

188 Ray Bradbury, *The Martian Chronicles* (Doubleday, 1950), first edition. From the collection of Barry Abrahams

189 Arthur C. Clarke, *The Sands of Mars* (Gnome Press, 1952), first U.S. edition. From the collection of Barry Abrahams

191 Robert A. Heinlein, *Stranger in a Strange Land* (Putnam, 1961) first edition. From the collection of Barry Abrahams

193 A. Bertram Chandler, *The Alternate Martians* (Ace Books, 1965), cover art by Jerome Podwil. Author's collection

195 Courtesy of SpaceX

196 *Mars Attacks, 1996. @ Warner Bros. / courtesy Everett Collection*

197 Topps® *Mars Attacks cards. Courtesy of The Topps Company, Inc.*

198 Kim Stanley Robinson, *Red Mars* (Bantam, 1993), first U.S. edition, first printing; cover art by Don Dixon. From the collection of Barry Abrahams

200 *The Martian*, starring Matt Damon, 2015. TM & copyright © 20th Century Fox Film Corp. All rights reserved / courtesy Everett Collection

202–203, 226 © Foster + Partners

205, 206, 212 Courtesy of SpaceX

221 NASA

225 © Foster + Partners

227 SEArch+ / Clouds AO.

232 Courtesy of SpaceLife Origin

235 Rick Guidice, NASA Ames Research Center, circa 1970s. Wikimedia Commons

236 NASA/JPL-Caltech/University of Arizona

238–239 Concept/design by John Spencer, artwork by Brian Cho. © 2019 Mars World Enterprises, Inc.

240 NASA

245 © Foster + Partners

254–255 NASA/JPL/University of Arizona

ACKNOWLEDGMENTS

When I set out to write this book, I knew a lot about early beliefs in Martians, but I didn't know a lot about Mars and the science surrounding it. History is one thing, but astrophysics and planetary science are entirely another. Fortunately, many brilliant scientists, engineers, and members of the space community took time away from their own Mars missions to share their knowledge.

Chrishma Singh-Dewera spoke to me before I knew much of anything and was kind enough to talk me through many aspects of Mars and its exploration over numerous phone calls, emails, and a full day tour of NASA JPL. The introductions to several of his co-workers is also greatly appreciated. John Spencer discussed many thoughts over the phone and at his home studio, and he graciously introduced me to others in the space community. Gil Levin and Pat Straat shared their remarkable experiences and perspectives, and ensured I captured it all correctly. Rachel Tillman generously provided notes, images, and further contacts. Ann Devereaux kindly welcomed my family into her JPL office and connected me to her colleagues. Lauren Amundson filled my research room at the Lowell Observatory with boxes of files, notes, and clippings, and then emailed a few more gigabytes' worth of images. And Kevin Schindler, Lowell's historian, spent his morning with me in that research room to share details and stories about Percival Lowell's life and mission.

So many others took time to chat on the phone or in person, email, and share documents to help me understand the many fascinating and complex programs, beliefs, and plans surrounding the Red Planet. These distinguished people are Matthias Biniok, Daniel Buckland, Chris Carberry, Ben Clark, Egbert Edelbroek, Ken Farley, Melissa Guzman, Ruth Hemmersbach, Jerry Hubbard, Alexander Layendecker, Gentry Lee, Pascal Lee, Jim Logan, Gioia Massa, Chris McKay, Ben Parker, Hoppy Price, Steve Squyres, Ashwin Vasavada, and Artemis Westenberg.

You've all made me a smarter and even more curious person, and for that I'm forever grateful. Thank you for your help and for the incredible and important work you do.

Others were helpful without being scientists or experts on Mars in any particular way. Thanks to Paul Guidry, Chinkara Singh-Guidry, Doug Latino, Grace Lynch, and Kevin Oliver for connecting me to amazing people they just happened to know or be related to; to Lee Speigel for sharing thoughts and information on UFOs; to Barry Abrahams for sending images of his expansive book collection; to Anne Archer at BT Heritage & Archives for scanning images from the Hugh Mansfield Robinson archive; to the staff at Grover's Mill Coffeehouse & Roastery for giving me access to memorabilia hanging on their walls; to Rabbi Scott Weiner for clarifying a few notes around Hebrew writing on Mars; to Warren Park for unknowingly putting me on the path that led me to Oomaruru; and to my parents, Bev and Paul Hartzman, who probably didn't realize it but gave me a name that many believe has its roots in the Roman god Mars, and also for reading an early draft of this book and pointing out pesky typos before they got to my copy editor.

None of these people would have been able to help me at all if not for my agent, Katie Boyle. Her enthusiasm and guidance for this project fueled it from the beginning and helped me shape it into a book for Quirk. I'm also grateful to my editor, Jhanteigh Kupihea, whose vision put me on a journey into places I never thought I'd go. Her notes throughout have been invaluable. And Mary Ellen Wilson's brilliant copyediting and Ryan Hayes's artful design have made this book better to read and beautiful to look at.

Finally, thank you to all my friends and co-workers who've listened to me yammer on about Mars and Martians for the past two years. Especially my wife and daughters, who've heard it more than anyone, helped me dig up articles, and accompanied me across the country for my research. Thank you Liz, Lela, and Scarlett for being the wonderful Earthlings you are.

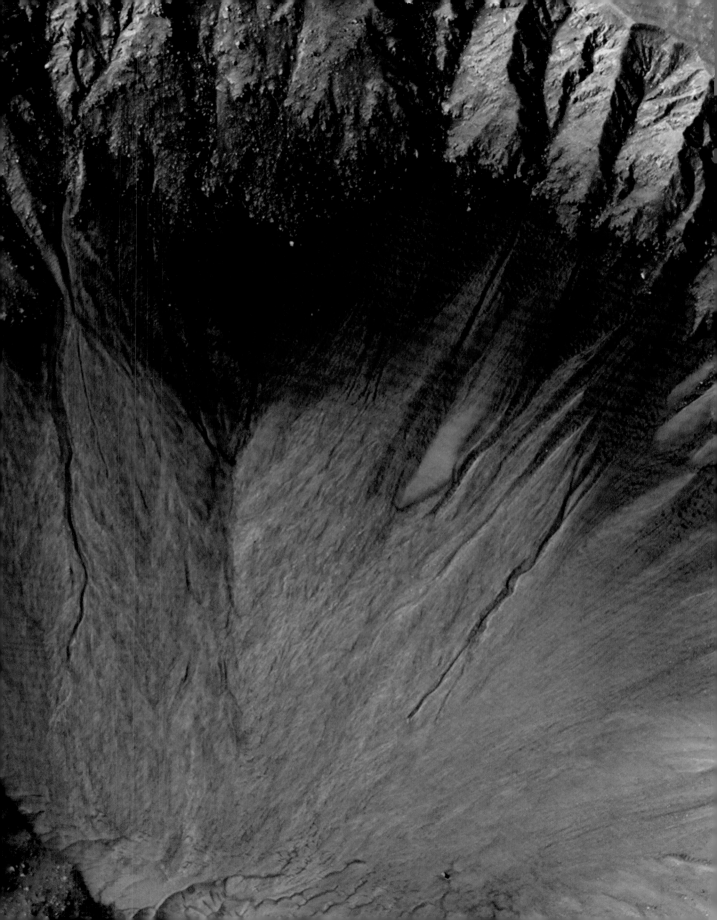